# Praise for *www.R U in Danger.net*

A must-read for all family members. Very insightful and thought-provoking. Parents, whether we like it or not, times have changed and we must adapt to what is happening with our children and their interaction with technology. We need to be involved in teaching our children about technology and not relying on them to teach themselves. Our children are facing a very different world. They are far more exposed to others than we realize. Many, sadly, are not thinking through the long term consequences for the information and images they are sharing. Take the time to sit with your children and learn the lessons this book has to offer.

~ Mike Piechota,
father of four and a forensic examiner
and police officer in Western Colorado

My main goal as a Youth Officer in Law Enforcement is to keep my community's children safe in town and on-line. It's hard to convince teens that what they do on-line or through texting can impact their lives for many years to come. The authors of *www.RU in Danger.net* use an original platform to not only present real-life scenarios to teens, but also test teens' abilities to recognize the dangers and consequences of common day occurrences in today's "sexting" and Internet world. The book is a must-have teaching aid for law enforcement, school systems, parents and teens everywhere.

~ Kristina Vetrano,
Police Officer

The chapters were eye-opening and made me think twice about what I do online.
~ Megan, 8th grader

A must-read and discuss book for all parents and their children. Each chapter follows the actions of a specific teen and illustrates how easily something seemingly innocent can quickly turn into a nightmare such as having contact with a would-be predator in an online chat room or cyber bullying on a social networking site.

~ Dana Sanderson,
Mother

This book showed me the reality of all the dangers that are on the Internet.
~ Maggie, 7th grader

Unfortunately, this kind of brutality is a reality to some kids today. I believe kids should be educated about how quickly things can spin out of their control, because they don't fully grasp the seriousness of the repercussions that will be incurred in today's modern cyber society. I want to share these stories right away with my class and hopefully enlighten our teenagers. I really appreciate the authentic style used in the fictional scenarios and I found myself putting faces to the kids in the stories . It's real and a very useful tool for teachers today.

The scenario in Chapter 12 seems like a very accurate depiction of what unfortunately happens in our schools. The Phoebe Prince case has had an impact on schools in Massachusetts. We now address our students and educate them in what bullying is and its effects on others. Everyone's behavior needs to change. Bullying can't be tolerated by anyone! As an educator, I want to help kids to understand that encouragement and compliments will be much more effective than insults and criticism.

I think these stories will be very effective in educating our kids about responsibility and repercussions that could last a lifetime. We live in a modern and very intolerant society. Laws have changed and educators and law enforcement are becoming very aware of the effect kids have on each other. I believe schools should provide the stories in this book to their students!

~ Steve Edwards,
Secondary Public School Teacher
- 16 years experience

As grandparents of kids who are starting to use technology for school projects and entertainment, this book is an essential tool in our grandparenting tool box. We don't want our grandchildren to fall into the kinds of safety traps the kids in the book's scenarios did, and this book educates them, and us, on how to avoid them and stay safe.

~ Bob and Margaret Rudenauer

An interesting — and important — read for any teen or tween in your household. What they will learn from each chapter will make a difference in the lives of today's tech-driven young adults. This book is going under the Christmas tree for my young teen daughters and their cousins!

~ Suzanne Cordatos,
Mom and High School Guidance Tutor

# www. R U in Danger.net

## Are you in danger? More than you know!

**Scott Driscoll**
**and**
**Laurie Gifford Adams**

Credits:
Cover Design: Bethany Thompson

Photographs: Little L's Photography
www.littlelsphotography.com

iUniverse, Inc.
New York Bloomington

**www. R U in Danger.net**
**Are you in danger? More than you know!**



*iUniverse books may be ordered through booksellers or by contacting:*

*iUniverse*
*1663 Liberty Drive*
*Bloomington, IN 47403*
*www.iuniverse.com*
*1-800-Authors (1-800-288-4677)*



*ISBN: 978-1-4502-6564-5 (pbk)*
*ISBN: 978-1-4502-6565-2 (ebk)*

*Printed in the United States of America*

*iUniverse rev. date: 11/15/10*

*To all of the Law Enforcement Officers who work tirelessly to protect children from predators*

# Scott's Acknowledgements

I have many people to thank and acknowledge for this project. It has been a very rewarding endeavor to take on, and I could not have done it without them.

First, my wife Bonnie, and kids Tyler and Amber. They have been my biggest supporters encouraging me every single step of the way. Their love and support have not only helped but been a huge motivation for me. I love you guys very much.

My partner in crime, Laurie Gifford Adams. When I first approached Laurie and presented my idea about this book and how I wanted to write it, she informed me that she needed to "let it ferment." Being the patient person that I am, I waited. I think it was only a matter of hours and she called me and told me she was in. From that moment on, it was fantastic working with her. She tells people that this book is my baby and she was lucky to be included. Let there be no doubt – if it were not for Laurie this book would still just be a thought in my mind. I have learned a great deal from her, and I know that this is just the first of many collaborative projects for us. (That's right, Laurie, you are stuck with me for a while.)

The Sanderson Family. Thank you for being our "testers" on this book and sharing your thoughts. Your opinions and input were extremely valuable and were greatly appreciated.

Little L's Photography. Laura, owner of Little L Photography, and I have been friends since we were kids. When I asked her to come up with photos for the book that would enhance the reader's experience, she came through with flying colors. Anytime I had a request for a certain idea she was right there for me. Thanks, Laura. www.littlelsphotography.com.

Dave Polochanin. For the time and energy you put into this as our editor. When Laurie and I would come running into your classroom with a thought, idea or proposal, you always took the time to listen and help us, even if we were completely throwing you off track from what you were doing. Thanks, Dave.

John Danaher. When I am all done with this law enforcement career and I look back, some of my best memories will be the cases that John and I worked together. John was working as an Assistant United States Attorney and I was a local officer assigned to a federal task force investigating crimes against children on the Internet when we first met. From the first meeting John and I worked very well together. We had some crazy cases and we were able to assure that some very bad people were no longer able to hurt children. When I first started online investigations, it was a new style of police work. On occasion, I would have a unique thought or idea about how the undercover investigations were going. I would call John and ask, "What do you think?" Sometimes, we would talk and a new game plan would develop. Sometimes he would make a simple suggestion that would fit perfectly. And there were a few times that my ideas made us both laugh and he would suggest not doing what I proposed. But no matter what I talked to John about, he always listened and respected my opinions and tactics. I became a better police officer working with him.

This book has been in my mind since my first Internet arrest in 2003. The reaction to that arrest was alarming, not the arrest itself, but how uninformed people were about what was happening on the web. Before that arrest and to this day, I try to educate anyone who will listen about the dangers of the World Wide Web and today's technology. People always ask me how I stay focused and keep emotions under control when you are talking and dealing with predators. It is not always easy but a couple of things stay in my mind.

First, my children. I always have a picture of them close by whether I am examining a computer for inappropriate material against kids or when I am posing as a child on-line while talking to potential pedophiles. My kids are always close to my mind and heart and are the reason I do it. My goal is to help protect kids from getting lured into a terrible world.

Also, the victims of these crimes help to keep me focused. Almost all of the victims of child pornography and sexual abuse did nothing wrong; some are still too young to even know the difference between right and wrong. They are innocent victims who have been through traumatic life-changing events and they need all of the support and help that can be given. Every time a child pornographic image is viewed or shared, that child is re-victimized and this isn't fair or acceptable.

The aforementioned reasons represent why I have dedicated my work in this book to the men and women in law enforcement who do the toughest job. Every aspect of law enforcement has its challenges and stresses. I dedicate this to the officer who sits on-line talking to the predator who wants to steal a child of his/her innocence. It is for the officer who has to look through hundreds of images of child pornography in an effort to make an arrest. It is for the officers who have had to see images, scenes, and videos that no person should have to examine and still maintain compassion and care to assist the victims. It is for the officers who put the safety of children first. These officers are true guardians over children and heroes in all of our lives.

# Laurie's Acknowledgements

First, and foremost, I thank my husband Jim and children Carrie Beth and Nick for their support throughout my writing endeavors and particularly on this book. Their insight and suggestions enabled us to make the fictional scenarios as realistic as possible.

Scott Driscoll. You pulled me out of my comfort zone to collaborate on this book. Even though I was hesitant at first, I can admit now it has been an incredible eye-opener as I've learned from your experiences. I've also enjoyed watching you come into your own in the writing arena. Congrats on finally seeing your idea come to fruition – or publication, if you prefer.

Next, I'd like to thank my critique partners Dave Polochanin, Suzanne Cordatos and Michelle Hacker. Your candidness and suggestions were greatly appreciated. Thanks for letting me throw this project at you in the midst of our fiction writing.

Bethany Thompson, our cover artist, deserves a great deal of credit for putting up with us for over two months while she created this cover. Her patience, diligence and artistic abilities are exemplary.

Zita Christian, friend, fellow author and host and producer of the cable access show, Page1. Thank you for believing in this project and its goal to protect children and families. We appreciate your support.

Finally, to my writing friends and fellow chapter members of Charter Oak Romance Writers. Even though this isn't a romance book, your enthusiasm for writing and encouragement on this project helped keep me on track during some of the more difficult chapters. The completion of this book is due in part to BISW.

# FOREWORD

It is an oft-repeated joke that most adults depend upon their children to help program and operate the many electronic devices that dominate our lives. It ceases to be a joke when we come to recognize that those devices are more than pathways to information and music. Those pathways, like all pathways, run in two directions. Our children, through the Internet, have access to the world, with all of the wonders and dangers that accompany such access.

How many of us feel completely comfortable in warning our children about the things we know so well: which foods are not healthy, which parts of town are not the safest, or which movies are not appropriate for them? How many of us, however, shy away from engaging with our children about the dangers that lurk in cyberspace? Why do we avoid talking to our children about such a critical issue? Is it because we just don't know enough about how it all works? Is it because we are too intimidated to open such a discussion with our children? Many of us are at least somewhat capable in working with our computers, but most of our children are far beyond "capable"; they are fluent with computers and the other, similar devices that constantly emerge in our ever-developing electronic world. Today's children have grown up with the Internet, and most of them know it just as well as they would know a foreign language, were they to spend their formative years in another country.

Scott Driscoll and Laurie Gifford Adams have provided us with the tools that we need to help our children. This book educates parents in an easily accessible format. It offers scenarios that provide multiple opportunities for parents and children to learn of the risks of the Internet and how to avoid them.

For twenty-one years, I served as an Assistant United States Attorney, and eventually as the United States Attorney, for the District of Connecticut. Thereafter, I served as Commissioner of the Connecticut Department of Public Safety. I worked with Scott Driscoll in both contexts, and was always deeply impressed with his abilities. In the 1990s, the United States Attorney's Office in Connecticut assembled a Crimes Against Children Working Group. It was comprised of representatives of the Federal Bureau of Investigation, the United States Customs Service, the United States Postal Service, state and federal prosecutors and investigators, and others, all dedicated to identifying, investigating, prosecuting and incarcerating those who would prey on our children.

Pedophiles follow familiar patterns. True pedophiles are obsessed with the idea of having sexual relations with children. Some collect images of child pornography. Some trade those images with others. Still others manufacture child pornography. And the most dangerous of all are those who act upon their impulses. Some pedophiles place themselves in positions of trust which give them access to children. Others troll the Internet, hoping to establish connections with naïve youngsters who, working their way through the confusion of their early teens, are susceptible to the blandishments of unseen people who, through the anonymity (and therefore apparent safety) of the Internet, tell those young people that they are attractive and desirable. Young teens who would not think of inviting strangers into their homes do exactly that, thousands of times each day, when they enter chat rooms or otherwise encounter people on the Internet.

Pedophiles spend their waking hours planning ways to get close to children, to earn their trust, and to ultimately exploit those children for their own purposes. Studies have shown that many pedophiles who are incarcerated have had hundreds of victims in the course of their "careers." One of the most dangerous and predatory groups of pedophiles are those who are known, within law enforcement, as "travelers." Travelers are not content to settle with viewing pornographic images, nor do they find sufficient satisfaction with cybersex. Instead, their goal in life is to establish an Internet relationship with a child and eventually convert that relationship into a sexual encounter.

Scott Driscoll, who for years has dedicated himself to capturing travelers, is one of the most accomplished law enforcement professionals I have ever encountered. He is able to pose online as a young teen, and he does it in a way that avoids potential legal pitfalls. He finds travelers, he skillfully avoids their traps, all of which are intended to distinguish law enforcement personnel from real children, and he prepares cases that, invariably, lead to convictions. When we consider that pedophiles often have multitudes of victims, the value of Scott's work is readily apparent. It is not possible to know how many children have been saved, over the years, due to the many cases he has solved.

Scott Driscoll and Laurie Gifford Adams are now reaching out, even farther, in order to help protect many more children – your children – from risks that are real but nonetheless avoidable. When your children were young, you protected them from hot stoves, sharp thorns, and many other obvious dangers. It is time to teach them to navigate, safely, the most powerful communications and information tool in history. This book will help you meet that goal.

John A. Danaher III

# PREFACE

Kids, how often do you allow strangers into the house while your parents are at work? Have you ever received a message of a sexual nature? Is there personal information about you and your family on the Internet that you are not aware of? If your family uses the Internet then you can probably answer yes to all of these questions.

One of the most exciting and widely used communication tools in our world today is the Internet. Technology is fantastic and changing every day. We use it for homework, research, social networking and even shopping. Most changes are awesome and make our lives easier and more enjoyable, but with every great change, dangers can follow if we're not smart.

Parents, our number one goal is to protect our children. We believe if they're home, they're safe. Before the world was introduced to the Internet that was generally true. But the Internet has changed our children's worlds and made them more susceptible to dangerous situations. Adolescents are particularly vulnerable because of their desire to grow up and be part of a social network of friends. Even with parents sitting one room away, children can be in danger if they're on a computer with Internet access.

Usually kids don't go on the Internet looking for trouble. Unfortunately, just by logging on, trouble can find them. Studies show that one out of every seven kids has received unsolicited sexually explicit material over the Internet. Another study indicates that the average age of first exposure to Internet pornography is eleven years old.

As an undercover Internet crimes investigator for many years, Scott witnessed the many avenues predators use to perpetrate Internet crimes. With this book, you'll learn how circumstances related to Internet use change rapidly and, rest assured, predators are keeping up. If you drop your guard, you can put information out to the world that shouldn't be in the public domain and can compromise your safety. This is why it is essential to be up to speed on Internet safety.

Our hope is that kids and parents will read this book together and that it accomplishes three goals. First, the format is intended to encourage open discussion between parents and children about what they have experienced already on the Internet and how to keep everyone safe. Secondly, when you participate in the ISSC (Internet Safety Savviness Challenge) section of this book, both children and parents need to be open-minded to what the other has to say and the experience or motivation that shaped those answers. Parents can gain new knowledge in this area, because undoubtedly your children may know more about this topic than you. You need to listen to them. Teens and preteens also need to keep their minds open and realize that some parents may not know the hottest trends in technology, but they are very knowledgeable about safety.

Finally, we hope the next time you go onto the Internet that you ask yourself, "Am I being safe?" If the answer is yes, use the Internet for all its benefits. If you ask that question and you pause, even for a split second, step away and think before you do anything on-line. If you are doubting your safety then you need to stop and reevaluate.

Safety awareness is every Internet user's responsibility. The purpose of this book is to make you aware of the dangers on the Internet and how to protect yourself and your family. By the time you finish this book you will be much savvier about the Internet and Internet safety. Prevention through knowledge is the best means to insure that you, your friends and loved ones don't become victims of an Internet crime.

# CONTENTS

# CHAPTER 1

# Andrew's Accidental Search

**Profile:**
6th grader
Lives with parents
Mother: Administrative Assistant to principal
Father: Carpenter
Siblings: none
Town demographics: mix of blue and white collar workers

"Punishment and discipline have changed throughout the course of history." Mr. Bailey, my 6th grade social studies teacher, wheeled and slapped a ruler against the desk. Every one of us in the class jumped. He bent slightly at the waist, narrowed his eyes and slowly scanned the room, making sure we were all paying attention.

Class is never boring with him because he always creates activities that make what he's teaching easy to understand. That's why when he planned the debate and divided our class into the two sides we were all so excited. Since I hope someday to be a lawyer, this activity was perfect for me.

We learned debate etiquette and vocabulary, including "telling points" and how those are the points that your opponent can't argue back against. More than anything I wanted to be the king of telling points. After preparing us for how to do a real debate, he gave us the topic.

***What's more effective: the way schools are run today or the way they were run a century ago?***

He chose the two teams, and I was totally psyched when I ended up on the side that was for how schools were run in the olden days. I like history, so that topic seemed more interesting. The directions were to use first person accounts along with books and the Internet as our supporting evidence. I knew I was all set with the first person account part.

At Thanksgiving dinner the week before, my grandparents talked about what school was like for them and even their parents. That was a really long time ago, and we joked about whether they used chisels and stone tablets to write with then. Between what they'd tell me and my Internet searches, I'd come up with a lot of good information for my team. I love winning, so I was determined to get the most and best information for this debate.

There were twenty minutes left in the period when Mr. Bailey instructed us to make a research plan with our group members. Each of us picked one aspect of old time schooling to find information about. My assignment was to make a list of punishments students received in the early $20^{th}$ century and what the punishments were for. This was going to be awesome!

I had Science Club for two hours after school, so it was almost dinner time before I started my research. First, I called Grandma Shay to set up a time to interview her and Gramps. Mom told me I could invite them to dinner the next night, which ended up being perfect for them. When I told Grandma about the debate and which side I was on, she was almost as excited as I was. She told me she'd even bring old pictures and report cards from when they were kids. I figured those had to be pretty old!

I hung up the phone and went right to the computer in the family room. With my debate notes worksheet in front of me, I clicked on the search engine and entered the topic of *school punishment in the 1900s*. Many options came up for websites to explore. One that looked interesting was about school in the 1900s. That seemed to be a good place to start, so I clicked on it. Immediately, I saw a page full of interesting information. I liked that the website was divided by decades starting with 1900. I could click on each decade and it told me what schools were like and how kids were punished.

Our library media specialist had taught us the importance of citing our sources, so I wrote down all of the information about the website. I had half a page of notes when I started getting suspicious about how good this site really was. I noticed that some of the sentences were written kind of funny. After clicking through a few more of the tabs, I discovered the problem. This wasn't a website created by an expert. It was actually created by elementary school kids for a class. Because of that, I wasn't sure how accurate the information would be. I didn't want to get in the middle of the debate and lose points because I'd used bad information.

I started my search again. This time I put in the keywords "history of discipline in schools". I scanned the possible websites then saw the one that looked like it would be perfect: *www.oldschooldiscipline.com*. I couldn't believe I might have found one site with all of the information I needed to score my telling points in the debate.

The website was set up in a cool way. On the home page there were photographs of old schoolhouses in different states. The directions said to click on a schoolhouse to get more information. I clicked on the red schoolhouse with the big bell on the top. The screen changed to an enlarged version of the picture. A heading called "Entrance Exam" popped up with a little bell to click on. The instructions said I had to correctly answer a question from the time period of the schoolhouse before I could enter. I loved it! I knew this was going to be a really cool website.

I clicked on the bell, and the question popped on the screen.

*Which U.S. president was a Rough Rider?*

I was given four choices: George Washington, Abraham Lincoln, Theodore Roosevelt and Woodrow Wilson. Being the history fan that I am, I knew about the Spanish-American War in 1898 when the United States helped fight for Cuba's independence from Spain. I smiled, proud of myself, and clicked on Theodore Roosevelt, the man who had created the cavalry regiment known as Rough Riders. The school bell rang and the front door swung open with a loud squeak.

This site kept getting more awesome. Inside the schoolhouse were pictures of four women who looked like old-time teachers. Their hair was wrapped up on top of their heads, and they had little glasses perched on the ends of their noses. Each teacher looked strict. I was glad none of my teachers looked like them. I had to pick a teacher for the school, so I clicked on Miss Snodgrass because I thought her name was funny.

Another message appeared on the screen.

*Miss Snodgrass caught little Ada Mae playing with the buttons on her dress instead of studying her spelling words. Miss Snodgrass wants to use Ada Mae as an example to teach all of the other girls in the class a lesson about playing with their clothing. What discipline should Miss Snodgrass use?*

It was fun playing this game, so I e-mailed the link to my group members so they could see what information I was getting. I figured they'd have as much fun with this website as I was having and also learn the same information so our team would be really strong in the debate. After clicking send, I returned to the website to explore more.

The next task was to decide what punishment Miss Snodgrass should give Ada Mae. I was given five choices.

1. *paddle – 2. knuckles rapped with a ruler – 3. stand in corner – 4. dunce cap – 5. write 100 times "I have been naughty"*

Each punishment listed had an icon next to it to click to make your choice. "Paddle" showed a little paddle that looked like a ping pong paddle. It swung back and forth and made a cracking sound like it was hitting something. Next to number two there was a ruler that repeatedly broke in two with a loud snap. The other three options didn't look as interesting.

I'd heard about some of these punishments and two of them were still sometimes used by one of the older teachers at our school, Mrs. Campbell. We joked that she had to be at least seventy. Kids talked about her making them stand in a corner for talking. I'd even heard of kids having to copy the same sentence a hundred times. I chuckled when I imagined her picture on this website.

I thought about Ava Mae as if she were a real person and wondered why she deserved to be disciplined. It made picking a punishment even harder. My grandparents had told stories about teachers paddling them in school, so I was curious about that. I was hesitant to click the icon, but then I stopped and laughed at myself. This was just a game that was teaching kids about old school discipline. Ada Mae wasn't even a real person. No one was going to get hurt, so I'd just check them all out.

I clicked on the swinging paddle and gasped when a picture I hadn't expected zoomed into focus. My heart hammered against my ribs. There was a picture of a naked woman bending over like she was waiting to be paddled. How did a picture like that get on a site for kids? I clicked on the home page icon and chose another old schoolhouse. I followed the same directions for that schoolhouse until it came to the discipline given. Each time when I clicked on discipline, another inappropriate picture filled the screen.

I was horrified and quickly closed out of the website. What would happen if someone found out I'd seen those pictures? I picked up the pencil and crossed this website off the list, relieved that I hadn't been in school when I clicked on it.

Then an even worse thought popped into my mind. I had forwarded the website link to all of the members of my group.

# I.S.S.C
# Internet Safety Savviness Challenge

This crossword puzzle will help you learn important terms for website searches.

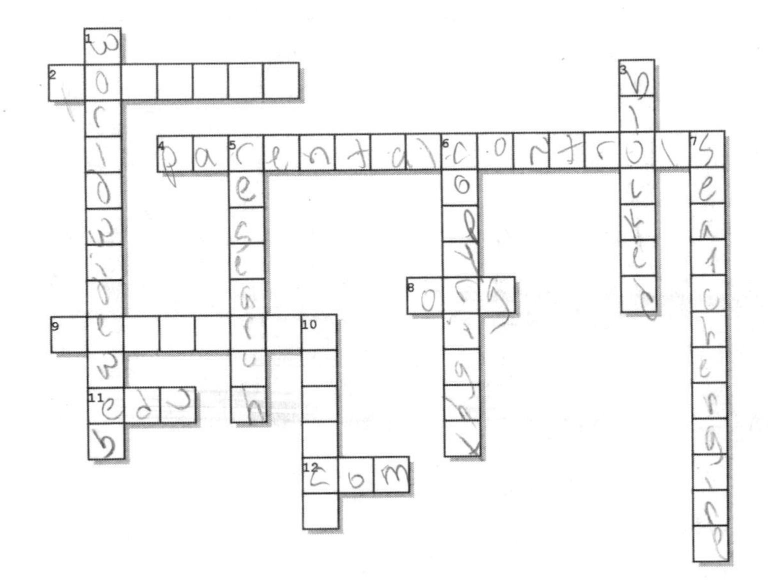

**Across**

2 term used when putting a message on website

4 settings adults should use to keep kids safe while on the Internet

8 represents a website for organizations

9 the word or words you use to start a search on the Internet

11 represents a website for educational systems

12 represents a website for a commercial business

**Down**

1 what www stands for

3 term used when software or settings prevent kids from going to a site

5 what a teacher will assign you to do/what the Internet can be used for

6 legal right to reproduce, publish, sell, or distribute matter

7 type of website that allows you to search for information

10 the settings parents should have all search sites set to

# What Can We Learn?

This situation shows just how fast things can get out of control on the Internet. Andrew is a great kid who loves school and tries his best in everything he does. Did he sit down at his computer and say, "Let's see how much trouble I can get into today?" Of course not, but that *is* what happened. The Internet offers an endless amount of legitimate web sites that can make our lives easier and more enjoyable. Unfortunately, the reality is that there are also many sites on the World Wide Web to which children should never be exposed, whether on purpose or accidentally. We can't ignore that these types of sites exist, because they do, and they will always be around. What we can do is learn how to prevent being exposed to them and how to deal with them if we accidentally come upon them.

Once Andrew discovered that the site he went to was not appropriate for his age he *almost* did the right thing. He navigated away from the page but did not leave the site; he went back and tried again. This was the wrong approach. Andrew should have immediately left that web site and notified his parents.

Right now kids are probably saying, "Yeah, right, tell my parents I saw an inappropriate picture. I'd get in so much trouble that it's better if I just ignore that site from now on."

Ignoring the site is a good approach, but parents need to be advised. Andrew could very easily explain to his parents what happened and how he did nothing wrong. Kids, parents will understand this if you explain it to them and be honest.

Let's walk through what could happen if Andrew does not tell his parents. He continues to work on the computer like nothing happened. He works for a while then decides to watch television. A short while later, his dad sits down at the computer and starts surfing the web for news and sports scores. He opens the web browser software and goes to the address line. His favorite sport team is the Orlando Magic. When he gets to the address line he types in "O". When he does this, a drop down screen could appear with recent sites that were visited beginning with "O". What do you think is the first thing he is going to see? *Oldschooldiscipline.com*. Not knowing what this site is, he will probably click on it and you know what he will find.

Parents, what will your approach be at this point? Would you say to your child, "Are you okay, and is there anything you would like to talk about?" or would it be, "WHAT DO YOU THINK YOU ARE DOING?"

Kids, this is why honesty is the best approach. If Andrew had told his parents immediately about his mistake, they all could have discussed this matter and dealt with it. If Andrew's father is the one who discovers the situation, now Andrew has to defend himself and prove it was an accident. That's not always the easiest thing to do. The point is, honesty is the best policy. If it was truly an accidental search then you've done nothing wrong. Unfortunately, these things can happen.

Besides the address line, there are numerous ways that parents can check what sites their families are viewing. You can check your browser's history. (See Appendix A at the back of the book for specific details regarding how to perform a history check on your type of computer.) This history check will tell you a lot about what sites your family members are visiting.

There are also ways to prevent children from getting to these sites accidentally. The first way would be by using parental control software programs. These programs can be set to block any site that has been deemed mature or inappropriate. These programs may not catch every bad site at first, but with regular updates to the software, they are a great tool for parents to use. No one program, software, or safety rule will prevent all bad things from happening, but each tool we use is another tool in our parenting tool box.

Another technique that can be used to help kids avoid these sites when searching the web is related to the settings parents can place on web site search engines. Most web site search engines have "search settings" or "preferences". In these options parents can set what search results can be shown. If you set the search option to "strict" or something similar, it will filter out inappropriate or "mature" sites. This means the children will not even see the sites or images so they will not be curious or tempted to check them out. Out of sight, out of mind is not a bad approach to these types of situations.

Let's revisit Andrew for a moment. Andrew thought he had a great web site that would help him and his classmates do the best they could in school. He forwarded the site to his classmates and then continued looking. Once he discovered the site was inappropriate, he realized he brought this to fellow teens' attention. Andrew should have immediately told his parents of the mistake.

Parents, what do you do at this point if you learn your child has made this mistake? Do you ignore it and go with the attitude that kids can't avoid inappropriate sites all the time, or do you do something? ***You must do something.***

Andrew's parents should contact the parents of all the students that Andrew notified and explain the situation. Again, let us make this point: Andrew did not do anything wrong, so everyone should work together to prevent it from happening again. Tell the parents that an accident occurred, and they need to be aware of how it happened and that their children should not visit the site. When parents communicate and network, only good things can happen. Parents need to communicate.

Should the school system be contacted? The teacher assigned this project and expected the students to do research on their own, and look what happened. The school system is not to blame, but parents notifying a teacher or administrator about what happened would be a smart thing to do. There is nothing the school would do discipline-wise, but the teacher who gave this assignment may re-think how to assign the research. Without knowing there was a problem, the teacher would continue to assign this research; however, if he/she was made aware of the potential problem, preventive measures could be taken in the future. The teacher could do research first and provide a suggested list of appropriate sites the students may use.

The administrators would want to know about the situation because most likely the students will be talking about the incident. The more information the school has, the easier it will be for them to deal with the aftermath in a safe and appropriate manner.

What about law enforcement? Should the police be called? Unfortunately, many of the sites that are mature or inappropriate are not violating any laws. Although most of us may say these sites should be taken down and permanently removed, law enforcement can not easily do that, and there are a few factors that affect this approach.

One, adult pornography is not illegal. Two, a lot of these sites, whether they are adult or child pornography, are hosted in places outside of the United States. In certain countries there are no, or very limited, laws preventing these sites. If law enforcement was contacted every time a pornography site was seen, that is all they would be investigating; however if you feel the images on a site are child pornography then law enforcement should be contacted.

It is important to understand that in these types of situations everyone involved has a responsibility. Children need to be honest and tell an adult and learn from this to be very careful when they are using web sites; not everything is what it appears or sounds like. Parents and other adults need to remember that the incident may in fact have been a complete accident, not a deliberate act, so remaining calm is a great start. Parents should then take the proper steps to avoid this type of incident again.

If a situation such as this does occur, use the incident as an opportunity to talk to each other about what was seen on the web, why it was inappropriate and why we need Internet safety. Open communication builds trust and trust is something that parents and children must have, especially when the Internet is involved.

# I.S.S.C
## Answer Key

## Andrew's Accidental Search

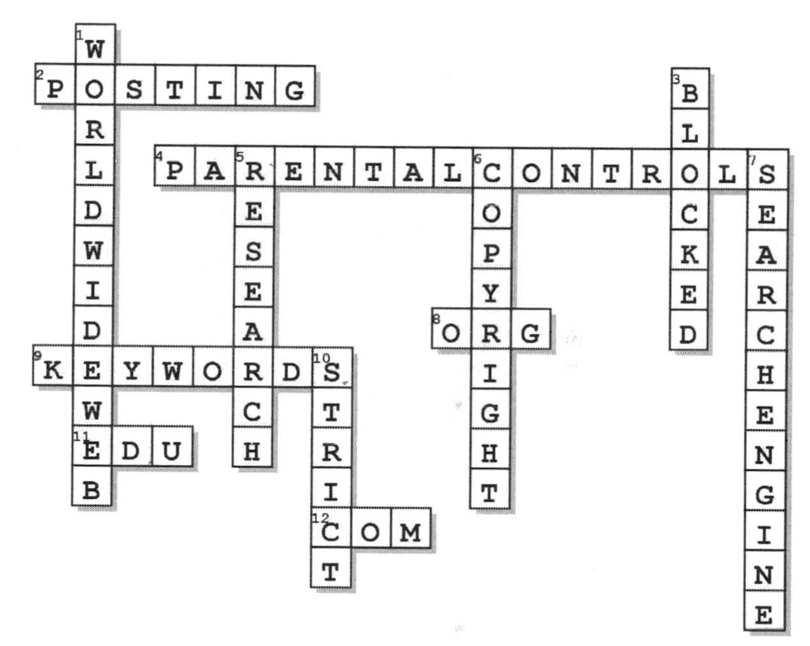

# CHAPTER 2

## Becca's Birthday

**Profile:**
14 years old, 8th grader
Activities: field hockey, piano, school newspaper
Living arrangement:
weekdays with Mom, a bank manager
weekends with Dad, an engineer, and stepmother, a teacher
Siblings: two older brothers and two older stepbrothers
Town demographics: city 214,000 residents

Being the youngest in a family has its advantages, but being the youngest and the only girl is even better. Even though my parents are divorced, Mom and Dad get along pretty well, and they actually still are parents together. That sounds kind of weird to most people, but my 14th birthday is a perfect example to illustrate my point.

Ever since I was twelve I'd been asking my parents for a laptop. I'd heard all of their reasons for me not having one: we had family computers that I could use any time I wanted (of course, that's if my parents weren't on them), computers were too expensive, or the worst one of all, I was too young. My brothers and stepbrothers had their own computers, so it seemed fair that I should have my own, too. They got theirs when they went to college, but like I told my parents, boys are different; girls mature earlier than boys and besides, times had changed since they were in high school. If I had my own laptop, like my friends, we'd all be able to work on homework and projects together. I'm not sure that they really believed my reasons, but those were my best arguments. I think the one that helped me the most was that I had joined the school newspaper club. After my parents saw my first article published they gave me more computer time, but sometimes we did have a conflict when they needed to be on, too.

After I'd put in my request for a laptop, at each gift-giving occasion I'd think maybe, just maybe, *this* would be the time I'd finally get it. When it didn't happen over and over, I finally gave up. Like my brothers, I would probably be 18 before I owned a laptop.

So, when my dad met my mom and me at the office supply store one night a week before my birthday, I was not expecting the surprise. They led me to the computer section, wished me a happy birthday and told me I had a certain amount I could spend on a laptop. I could pick out any one I wanted within that budget. After looking over every display the store had, I ended up with a pretty sweet computer that had everything on it that I wanted. I couldn't have it until my birthday, but I didn't care. This would be the best birthday ever.

Since my birthday fell on a Saturday, my dad and stepmom told me I could have a sleepover with my best friends Maddie and Corrina. My family waited to open presents until after they arrived. I opened the gifts from Maddie and Corrina first. Maddie gave me the I-pod case I'd been dying to get and a signed hard cover copy of the newest book by my favorite author.

"She was signing copies at Barnes and Noble when I was in New York City," Maddie explained. "I waited in line for over an hour because I knew you'd love it." She pulled a photograph out of her satchel. "And I took a picture of her signing it." She handed me the picture. "This can go in the book, too."

"Awesome! Thanks," I said, giving her a hug before I tucked the photo in the book.

Next I picked up Corrina's gift. She gave me a DVD of our favorite band and a sterling charm of a field hockey stick for my bracelet.

"Sorry the DVD's not signed," Corrina said. She gave me a mock sad face, and we laughed. It was unlikely we'd ever even get to hear the band live, let alone get an autograph.

They were great gifts, but I couldn't wait for Maddie and Corrina to see the gift from my parents. All of my friends already had their own laptops, but theirs were older and didn't have some of the built-in features like the video camera and microphone.

"O-M-G!" Maddie squealed when I pulled the paper off the box. "Is that really a laptop?"

Excitement bubbled inside me. "I picked it out last week. I wanted you to be surprised." I slid the computer from the box, and Corrina reached over and petted it, making us laugh. "It's not a dog." I pulled it out of her reach. "You can't pet it."

Mom and my stepmom disappeared into the kitchen then a moment later they came back out singing "Happy birthday". My mom carried a double chocolate cake with chocolate frosting while my stepmom carried the fudge ripple ice cream. This was the sweetest birthday ever, in every way.

When we finished dessert, Maddie, Corrina and I were excused and the real partying began. We took my gifts to my room and set up the laptop on my desk. It looked so good there, just the way I'd imagined. The night we'd bought the computer my dad networked it with the Internet and then put it back in the box, so now I was all set to use it. We went on-line and did some instant messaging with friends from school, surfed the web to find information on the band we liked, then left the computer for a while.

My favorite thing about sleepovers is that you hardly ever really sleep. We gave each other funny manicures and pedicures, painting our fingers and toes wild colors that we'd take off the next day. We played cards and filled in some Madlibs, but when we got too loud after midnight my dad knocked on the door and asked us to quiet down so the rest of them could sleep. Of course, that made us even gigglier, but we tried to settle in to watch a movie on my new laptop to quiet down. We were a little more than halfway through it when we started talking about school.

We were lying on sleeping bags on the floor. Maddie rolled over to stare at the school newspaper clippings above my desk.

"Do you ever wonder what happened to Joey Marshall after he moved?" she asked. Corrina and I also turned to stare at the newspaper page I had tacked there. Joey, the pitcher for our school baseball team, was at the top of every junior high girl's dream list. He was as nice as he was cute, and at the beginning of the school year I got to interview him. Shortly after, his family moved.

Corrina sat up. "Justine said he lives in Texas now."

"How would she know?" Maddie asked. "I'm sure he wouldn't keep in touch with her."

"She looked him up on-line," Corrina answered. "She said he has a profile on *FriendPost.*"

We looked at each other and gave the *Why not?* look then scrambled off the sleeping bags to go to my desk. I opened up the *FriendPost* social networking site and typed in Joey's name. Sure enough, there he was. We all sighed at once, then giggled some more. We clicked through as much of his page as his settings allowed. I pretended I was going to click on the button to invite him to join my network. Maddie grabbed my hand and tried to make my finger click the mouse. We wrestled over that for a minute, laughing about whether Joey would even remember who I was. Finally, I won and I closed out the page before the search went anywhere.

"Why aren't there good stories like that in the school newspaper anymore?" Corrina asked. She sat Indian style on the sleeping bags.

Maddie flopped back on my bed and stared at the ceiling. "Yeah, the stories you guys do are so lame. No one even reads them."

I turned in the chair and scowled. "Lots of people read them."

Maddie rolled her eyes at me. "Like who?"

"The teachers. They all comment after an issue comes out."

"They're the only ones, then. You need to write about something interesting and different. You know, write about something kids want to know about."

I hated it when the kids at school put down the stories in the newspaper because we worked hard on them. "If you know so much, why don't you give me a story idea," I challenged.

Maddie tapped her ridiculously painted fingernails against her lips. "I – um –"

It gave me some satisfaction to hear her stuck. I'd love more than anything to have a "scoop", as Mrs. Turkington, our advisor, called a really good story that no one else had reported on. Maddie was struggling to give me ideas when Corrina spoke up.

"Write about video chat games."

Simultaneously, Maddie and I said, "What are those?"

"I don't really know, but I saw a story on TV this morning about a video chat game on-line. It's called *Stop On Me*. It hooks you up with random people."

Corrina had my attention, and by the way Maddie suddenly sat up, she obviously had hers, too. "I don't get it," Maddie said.

"I don't know how it works, either, except you use a webcam and microphone and you can see and talk to people who are signed onto it at the same time."

I tried to picture talking to strangers on the computer. What would we talk about? "It sounds kind of weird to me."

"The reporter on the show was hooked up to it, and the network kept blocking out pictures on the screen while she was reporting. She was saying things like, 'Oh my' and 'Censor that', and stuff like that," Corrina said. "She made it sound like it was kind of a risky thing."

"That's dumb. How could it be risky if you don't even know the people?" Maddie asked.

Corrina shrugged. "I dunno. Some of the images that came up of people on there looked like kids our age."

"Is it free?" I asked.

"I think so," Corrina answered. "The reporter logged on in front of the camera during the show. I didn't see her put in payment information."

Maddie launched off the bed. "If it's free then it can't be bad. We should try it." She was right behind me in a second.

"Try what?"

She leaned over me and started typing. "Duh! *Stop On Me*. Let's see what it is." She clicked "go".

"I don't know."

"Oh, come on; it's a game. Corrina just said they talked about it on TV this morning, and it's free. I want to see what the big deal is."

As soon as she said that, the screen filled with pictures of men and women. The pictures changed every few seconds. A message scrolled across the screen welcoming us to *Stop on Me* with a slogan that said *Your dream mate is just a click away*. A hand appeared in the corner pointing to a box that said "Activate camera".

Maddie glanced around my desk. "We need a webcam."

"It's built in." I pointed to the little dot in the casing just above the screen. "Right there."

"Cool." Corrina moved next to us. "How do we activate it?"

"Um -" I searched my computer until we found the command. With one click the camera and microphone were on.

A list of rules appeared on the screen, and Maddie read them out loud. "*Players must be at least 16 years of age to play. All players must wear clothes.* What?" She threw her head back and laughed.

"Shhh!" I scolded. "You'll wake up my dad and stepmom."

"Are they serious? That's a rule? All players must wear clothes?" She covered her mouth with her hand to stifle her laugh, but she looked so funny that Corrina and I couldn't help but start laughing, too. Tears streamed down our faces from trying so hard not to laugh loudly. When we finally calmed down, I looked at the rules again.

"We can't play."

"Why not?" Corrina asked.

"Duh! We're not 16." I wondered why the obvious wasn't so obvious to them.

"Who cares?" Maddie said. "We won't play. We're just looking to see what it is."

"I guess that's true," I said, not totally convinced.

Another box appeared on the screen. *Click here to connect to game.*

The three of us looked at each other and shrugged. By now, I was pretty curious about this, too. The screen filled and then there were two screens, one on top of the other. The top screen showed a middle-aged guy without a shirt on playing guitar and singing a really strange song. The bottom screen showed us, well, all of me and only part of Maddie and Corrina because I was the one directly in front of the screen. To the right was a text box where we could write messages. Maddie grabbed the mouse and clicked the "next" button. A new guy, who looked like he might be 20, sat in front of his computer.

"Hey," he said into his camera, "nice pajamas."

We were so startled to have him talk to us that we screamed and jumped back from the desk. He laughed and suddenly he was gone and a woman was on the screen. She tossed her blonde hair back like she was a model, but she was missing several teeth in front. The next woman winked at the camera and blew a kiss toward us.

"Eww," Maddie cried and quickly clicked the "next" button. A good-looking teenaged boy came on the screen. "This is more like it."

The kid smiled back from his camera. "Oooh, a threesome. Wanna chat, girls?" He moved his face in closer to the screen. Like before, all three of us moved back.

"Uh, not right now," Maddie said, then she clicked next. Maddie moved away from the computer. "You know what? If hot guys like that can see me, I don't want to be in my pajamas. I'm getting dressed."

"What? We can't stay on this website," I said.

"Don't be a baby, Becca. You're 14 now. Of course you can be on this website. It's just a bunch of random people."

I glanced back at the screen. It had changed to someone else, just like it did every few seconds, even without us doing anything.

"I'm in," Corrina said.

"Come on." Maddie practically dragged me out of my chair. "Let's get nice clothes on if we're going to meet guys."

"We're not meeting guys," I protested, but it didn't matter. She was in my closet pulling out clothes. After a few minutes she'd picked out an outfit for each of us. We stood with our backs to each other as we yanked off our pajamas and put on the regular clothes. Corrina was a little bigger than us, but she managed to get one of my nice shirts on. We heard a whistle from the computer and turned back to see a wrinkly, gray-haired guy watching us from the screen. We'd forgotten about the webcam when we were changing. This time Corrina screamed, and I threw my hand over her mouth.

"Shhh!"

All three of us dropped onto the floor out of the webcam shot and froze, listening for movement from my dad's room. After a minute we decided we were safe. My lungs burned, and I realized I was holding my breath. I let it out slowly and stood up, moving to the other side of my bed so I could finish buttoning my shirt away from the camera.

When Maddie and Corrina were ready, we went back to the computer. We arranged ourselves so all three of us could be seen. The screen changed pretty fast with new shots of people every few seconds. We came to some girls, who looked like they were older than us, partying and holding up beers to the camera. One girl attempted to say something, but her words were slurred and her head rolled around like she couldn't control it. Behind them, a couple of guys were flopped on a couch like they were passed out.

"Yeah, that looks like fun," I said sarcastically.

"It's a blast!" the brunette girl said.

I looked at Maddie and Corrina. "Oh, my gosh! She can hear us!"

"Of course she can, you dope. We have the microphone on," Maddie said.

She reached past me and hit the mouse to advance to the next screen. A guy in a hoodie stared at us. We could barely see his face but a cigarette hung from his lips. Again we clicked.

This time a guy who might have been in his early twenties held a sign that said, *Flirt with me.*

Corrina leaned closer to the camera so just one eye showed. She winked slowly and said, "Hi, honey."

The guy laughed. "Cute. Got anything else to show me?"

She stood and turned her backside toward the camera then shook her behind at him.

"Nice," he said, dragging out the word. "Keep going."

"Corrina!" I whispered, "knock it off." I clicked "next" before she could do anything else.

"What?" she said, laughing. "It doesn't matter. He doesn't know who I am. I'll never see him again."

We clicked several times quickly. A teenaged girl appeared on one click. She lifted her shirt to the camera, showing us way more than anyone should see. A few clicks later we came to a man who was wearing only underwear. "That's gross," I said.

Maddie hit my shoulder. "Chill out."

"What if we see someone we know?"

"Like who? Mr. Wolfe?" Maddie asked, referring to our principal.

Corrina snorted. "Do you think he plays this game when his wife is sleeping?" she asked. That started us laughing again.

Now we were in silly mode and it was all downhill from there. Everything we saw was funny to us. We clicked *next* again and again. It was mostly older men, and some of them were doing very inappropriate things.

"O-M-G!" Maddie yelled.

"Shhh!" Corrina and I said at once.

"I'm closing out of this," I said.

Maddie swiped my hand out of the way. "No! Not yet."

"Some of these people are disgusting," I said.

"Just a few more," Maddie said, "then I promise we'll turn it off. What if we come to some cute guy?"

"Whoop-ti-doo," I said. "It's not like we'll ever meet him."

She kept clicking next. I turned from the screen but listened as they commented on some of the people they saw.

"He's naked!" Corrina suddenly shouted.

Without realizing what I was doing, I turned to look at the screen. There was a guy, who was probably in his twenties, sitting on a blue bean bag chair totally nude. I had had enough.

I reached for the mouse. "Click next." But Maddie moved the mouse out of my reach.

"No way," she whispered, wiggling her eyebrows. "He's cute."

"Maddie! Come on. He's naked." Now Corrina blocked me from the mouse, too.

The guy's voice came through the speakers. "Naked, you say? That would be *nuddy* to us." He had a heavy accent like I'd heard in movies.

"Huh?" Corrina and Maddie said at once.

"It's our word for naked." His voice was deep, making him sound older than my quick glance made me think he was.

Corrina turned toward me. "He's gorgeous. Why aren't you looking?"

"Corrina, he's nude!" My voice came out all tight and squeaky. I couldn't believe this was happening. If my dad or stepmom walked in, we were dead!

"You sheilas 'avin' a rage?" he asked.

"A what?" Maddie asked. I looked at the box on the screen that showed our faces then backed away. I didn't want him to see me.

"Oh, I get it," he said. "You sheilas must be in the States. I recognize yer accent." We looked at each other and tried to hold back giggles. Even though I was uncomfortable, I had to admit it was pretty funny that he thought *we* had accents.

"We're from Massachusetts," Maddie said, and I immediately popped her in the arm. She looked at me and I gave her my best "you're an idiot" look.

"This the first time you sheilas 'ave been on 'ere?"

"Sheilas?" Maddie asked. "Why do you keep calling us sheilas?"

"A sheila is a woman 'ere where I'm from," he said.

Corrina leaned toward the camera so only she could be seen. "Yeah. And where *are* you from?" She made her voice sound kind of sexy, something I'd only heard her do when we were goofing around and making fun of people in movies.

"I'm from Down Under."

Corrina's face lit up. "Australia?" She turned to Maddie and me and whispered, "Australia! We're talking to some guy from Australia."

"Duh!" Maddie said, shoving Corrina over so they were both in front of the webcam again. "Becca and I have perfectly good hearing, too."

"Oh, so you 'ave names. Becca is one. Who are the other two?"

My heart started beating fast. I tapped on Maddie and Corrina and shook my head "no", but Corrina told him anyway.

"I'm Corrina."

"Hey, Corrina, yer a pretty one," he said.

Maddie flipped her hair back, winked and said, "I'm Maddie."

"Well, well, darlin'," he said. "I think I've hit the jackpot here with you three."

This situation was getting worse by the second. "We have to turn this off," I hissed, but they ignored me. This was all too weird and dangerous.

"Do you always play this game without clothes on?" Corrina asked, then she and Maddie looked at each other and giggled.

I grabbed my head in my hands, afraid my brain would burst from the blood pulsing through it. Suddenly I had an idea for a way to end this since they weren't listening to me. I leaned down and yanked the plug from the outlet. But I'd forgotten about the battery.

"Nuddy is the best way to play this game," he said. "In fact, it's more fun if you sheilas show me what you've got. At least show me some grundies."

"Some what?" Maddie asked.

He laughed. "You Seppos. I 'ave to spell out everything."
"Seppo?" Maddie asked.

"That means Americans," he explained.

I couldn't understand how he could sit there naked and carry on a conversation like there was nothing weird about it. And worse, that Corrina and Maddie were going along with it.

"Okay, then what are grundies?"

"Underwear. Panties. However you want to refer to them."

"Oh, like these?" Corrina asked then suddenly pulled down her shorts. "What do you think of the stripes?"

"Corrina!" I yelled.

She and Maddie laughed really loud. My father and stepmother *had* to hear them. This was horrible.

"Ace!" the guy said, and he laughed, too. I didn't even want to know what "ace" meant, but he sounded happy about what he'd seen. "Okay, that's a start. Now, 'ow 'bout a little more. Maybe some skin. You're seein' mine."

"You want skin?" Corrina asked, her voice flirty.

She and Maddie were out of control, and I was really scared.

"No, Corrina. Turn this off now. This is bad." I tried again to shove past them to get to the laptop, but they pushed together so I couldn't get near it. I was afraid we'd knock it off the desk if we kept struggling.

"Who's the knocker?" he asked. "She's not as fun as you two sheilas. 'ow 'bout that skin you promised?"

I was shocked when Corrina lifted her shirt to the webcam, but not as shocked as when my bedroom door flew open at the same time.

"Becca?" My stepmother poked her head in the room, and I could see my dad standing off to the side in the hall trying not to look in. "You girls need to quiet down."

I knew the second she noticed the screen on the laptop because her eyes shot open wide and her jaw dropped. "Dave, you better come in here," she said to my father.

A lump bigger than a watermelon slammed up into my throat, making it hard to breathe. The best birthday ever took a sudden nose dive.

# ISSC
# Internet Safety Savviness Challenge

Here are five mistakes that were made in this situation. Below the mistakes are five alternative actions that could have been taken to prevent these safety issues. Match the "mistake" to the "alternative" that would have prevented the negative outcomes.

**MISTAKE:**

1. ____ Maddie suggested they go to the game site Corrina had heard about on TV, but Becca was uncomfortable with this.

2. ____ Corrina flirted with the strange man in one of the chats.

3. ____ The girls went on to an age-restricted website.

4. ____ The girls discovered people displaying inappropriate behaviors on the game site.

5. ____ Becca received the laptop as a birthday gift.

**ALTERNATIVE ACTION TO PREVENT A SAFETY ISSUE:**

A. Becca's father should put on parental controls.

B. Turn off the site and inform the parents.

C. Becca could have used her parents as the "bad guys" and told her friends she's not allowed to go to unknown sites.

D. Becca's parents should have shared their restrictions and expectations with all three girls before they were allowed to use the computer alone.

E. Never communicate with someone on-line whom you don't know.

# What Can We Learn?

There is a lot we can learn from in this situation. Becca tried to do the right thing on a couple occasions, but overall a lot of poor choices were made. Let's walk through this chapter in chronological order and see what happened and how you can make sure this never happens to you.

> ***When we finished dessert, Maddie, Corrina and I were excused and the real partying began. We took my gifts to my room and set up the laptop on my desk. It looked so good there.***

If a simple step had been taken by the parents, none of this would have happened. Computers belong in a common area of the home, not in a child's bedroom. Would the girls have experimented with this game if they were in the dining room with Becca's dad and step mom in the next open room or same room? Of course not, but when that bedroom door is shut and the world is accessed with the Internet, anything could happen.

Let's imagine that the computer was in a common area and Maddie and Corrina still wanted to try to get on this site. When Becca refused and tried to get them to stop, if she raised her voice or told the girls to stop, her parents would have heard it and been able to assist. In her room she tried, but there was no adult within hearing range to help, and look what happened.

> ***"I don't know, but I saw a story on TV this morning about a video chat game on-line. It's called Stop On Me. It hooks you up with random people."***

When we don't know what something is but we have heard about it, yes we are curious and want to check it out; there is nothing wrong with that. You can do research on it or ask someone who may know more. With today's technology we don't always research, we just jump right in and learn as we go. This is not always a safe method. "It hooks you up with random people" is a nice way to say "Come meet and talk to complete strangers". We all know that is not safe, but while sitting in our homes do we think about stranger safety and awareness? Probably not. These style games/chat sites invite trouble by connecting strangers.

***"The reporter on the show was hooked up to it, and the network kept blocking out pictures on the screen while she was reporting. She was saying things like, 'Oh my' and 'Censor that', and stuff like that," Corrina said. "She made it sound like it was kind of a risky thing."***

If a news report has to block out pictures, the directive "censor that" was used, and if it sounds like a risky thing, DON'T DO IT! If the girls had paid attention to these simple warning signs they would still be in the room watching a movie and having fun. Again, curiosity played a big part in this, but kids have to realize when things get "blocked out" there is a reason. There are things that a child should not be exposed to regardless of how mature they think they are.

***Your dream mate is just a click away...***

A place for kids? No. Enough said.

***Players must be at least 16 years of age to play. All players must wear clothes.***

The only rules this game site has is you have to be 16 and be dressed. These are obvious warning signs. First, no one enforces these rules. As we saw with the girls, things happen so fast, how can anyone supervise these games? There are no age verifications and no special software to assure someone is wearing clothes. The girls even reacted by laughing, nervous laughter, but laughter just the same. That nervousness again should have been a warning to not do what they were about to do, but they ignored it.

Look at this situation in a different way. If you have a birthday party or a get-together at your house, are these the rules you write on your invitation: You can come to my house for a party but please wear clothes. If you show up naked we will not let you in? Sounds crazy in "real life", doesn't it? It is just as crazy on the Internet.

***The screen filled and then there were two screens, one on top of the other. The top screen showed a middle-aged guy without a shirt on playing guitar and singing a really strange song. The bottom screen showed us, well, all of me and only part of Maddie and Corrina because I was the one directly in front of the screen.***

A complete stranger, an older man, can see and hear the girls. It does not matter if this person is half way around the world or down the street. It is a strange older male in Becca's room.

***"If hot guys like that can see me I don't want to be in my pajamas. I'm getting dressed."***

Once you have to dress up to use the Internet so strangers get a better look at you, you should not be on the Internet. If a stranger drove by your yard as you were playing in it and said, "You're cute", would you tell him to hang on while you go get nicer clothes on? No. You would get away saying, "What a creep!" The same rule should apply in the cyber world that applies in the "real world".

***We heard a whistle from the computer and turned back to see a wrinkly, gray-haired guy watching us from the screen. We'd forgotten about the webcam when we changed.***

A complete stranger was just allowed to watch three teenaged girls change their clothes. The person who whistled did not break into the house, did not break any laws and did not pressure the girls to do anything. They opened themselves up to the world, and this was a consequence of their actions. If the person who whistled was recording them, there is now a permanent record of these girls changing their clothes. Creepy thought, isn't it? But it happens.

***It was mostly older men who showed up on the screen, and some of them were doing really inappropriate things.***

Again, is this a place for children?

***I leaned down and yanked the plug from the outlet. But I'd forgotten about the battery.***

Becca tried to do the right thing here but was not having any luck and the other girls were not listening. At this point, Becca needed to be more assertive. She should have been more aggressive about getting the mouse so she could exit out, close the laptop or go get an adult. Now some may say this is tattling, but it is not. Becca gave her friends every opportunity to do the right thing and, when they did not, she should have stood up for herself and gotten help. Protecting you and your friends is not tattling; it is the mature and right thing to do.

***"Underwear. Panties. However you want to refer to them."***

***"Oh, like these?" Corrina asked then suddenly pulled down her shorts. "What do you think of the stripes?"***

If you would not do something in the real world, don't do it in the cyber world, either. Corrina would never do this to a stranger walking down the street, but here, in Becca's room, she gets "braver" and "dumber".

***I was shocked when Corrina suddenly lifted her shirt to the webcam.***

It is safe to say that Corrina has crossed a line that never should be crossed, and she can not take back her actions. She just exposed herself to a stranger who claims to be from another country. What if he is actually from Corrina's home town? What if he sees her at the grocery store with her parents the next day? What if he works with Corrina's father? There are a lot of "what ifs" that could be damaging.

Sextortion is a new word that has emerged in the cyber world, and sextortion is dangerous. People are using video chat sites to groom children into exposing themselves to the camera. Once they do this, the "bad guy" captures the image and then uses it against the kids. The kids are told that if they do not do what they are told, the image will be posted on the web for everyone to see, they will send the picture to everyone in their school or the bad guy will find the child's parents and send the picture to them. The children will be told to pose for more provocative pictures or perform "adult" style acts either live on camera or make a video and send it. It is extortion, just not for money, and the victims are children.

These video chat/games are exposing users to strangers from all over the world, and people are not thinking about safety or consequences of their actions. Things could have been different by taking these safety steps:

1. As stated earlier, the computer should have been located in a common/public area of the house.
2. Parental controls should be set to prevent these types of sites and games. (See Appendix B at the back of the book to learn how to set these controls to a safer mode.)
3. The parents should have rules established with Becca and her friends before any Internet use.
4. Becca should have followed her gut instincts that something was not right and stopped her friends before they ever became involved with the game.

Kids do not belong on these sites or games. Most sites say 16 is an okay age to participate, but the reality is different. No one under 18 should be on video chats. What starts off innocently, just chatting, can turn dangerous and illegal.

Let's say Corrina did more than just show her top and the guy from Australia recorded. Besides the possibility of "sextortion", two things could happen. The first is the video could end up on a website for the whole world to see. Secondly, if the video was deemed child pornography, and this man was found with it, he would face serious criminal charges and Corrina would be involved. The possibility of facing serious charges and consequences for losing control and not thinking is not worth the temporary perceived fun.

# ISSC Answer Key

1. C
2. E
3. D
4. B
5. A

# CHAPTER 3

# Troy's Troubles

**Profile:**
9th grader
High honor roll student
guitarist in a teen band
Parents: Restaurant Owners
Siblings: Brett - 20, Ben - 17
Town demographics: majority middle class - 23,000 residents

Thanks to my grandparents, I'm now a list guy. I keep lists for everything in a journal they gave me when I turned 10. I didn't know what to use it for because I didn't want it to be like some girl's diary. So one time after my dad, brothers and I went to a car show, I used the journal to list the cars and trucks we'd seen that I thought were cool. After that, I decided to use the journal for random lists.

Each list gets its own page: states I've visited, movies I've seen, my favorite bands, Yankees players and their stats, and music I want for my mp3. Actually, the "music I want" page has turned into five pages, because it seems like there's always a long list of songs I want. It was that list that got me into trouble.

For my 15$^{th}$ birthday my aunt and uncle gave me a gift card so I could download twenty-five songs from an Internet music store. I was psyched and bummed at the same time. It was a great gift, and I really appreciated it, but I had over a hundred songs listed in my journal. Trying to decide which 25 I would buy wasn't easy.

Four of my friends and I had started a band when we were thirteen. I played lead electric guitar. We usually did covers and didn't have music for most of the songs. We listened to songs over and over to pick out the notes. Over the last two years, we'd gotten really good at pulling out our parts from the songs. When we practiced it didn't take us long to put all of our musical parts together and sound really good. We'd even been hired to play for a couple of school dances and some bar mitzvahs and birthday parties. That was totally cool!

Dan and Marc, two of my friends who are in the band, came over. I showed them my song list and told them I had the gift card so we could pick out 25 songs. They pulled up two chairs on either side of me at the computer, and we opened the website for the music store.

My brother Ben, who's a senior in high school, came into the family room when we were going through my lists. He didn't say anything, which is incredible for him because he usually busts on my friends and me any chance he gets.

Part of the reason for this is that it bugs Ben that I make the high honor roll at school and most times he doesn't even make the honor roll. He's smart enough. As my parents say, he just has a bad case of "senioritis" and doesn't hit the books. I think he likes giving me a hard time about anything because he thinks it's his job as my older brother. Most times I ignore him when he's being jerky, but when he flopped onto the sofa and kicked his feet up on the arm like he was Mr. Cool, I figured we were in for it.

I stopped looking at my list and turned toward him. "What's up?"

"Nothing."

"What do you want?" I snapped. I couldn't see any reason for him to be in the room. He hadn't even turned on the television

He stretched and put his hands behind his head. "I came to offer a suggestion, but if you're not interested, that's fine."

I looked at Dan and Marc. Ben wanted to help us? "We don't need your help. We're picking out the songs I'm going to buy with my gift card."

He gave me a look like I was the biggest moron walking the earth. "Dude, I know. That's why I'm offering to help."

"You going to buy some songs for me?"

He snorted and shook his head. "That's a major waste of precious monetary resources."

Dan rolled his eyes, but Marc nodded like he agreed with Ben.

I pressed back into the overstuffed chair and tried to look as casual as Ben. "Okay, I'll bite."

"There's a site where you can download whatever you want, and it won't cost you a penny."

I looked at him skeptically. "Free? Mom and Dad made you take the downloading software off the computer last year. Remember?"

"This is different. This is a totally safe and untraceable site."

Marc suddenly sat up. "Cool! What is it?"

I whipped around toward him. "There's got to be some catch, Marc."

"Dude!" Ben said, sitting up. "You're acting like a little kid. I'm telling you, no one even knows about it so the *Internet police* aren't going to be looking for it." I hated his mocking tone, but I didn't want to get in trouble.

Dan laughed. "The Internet police? Troy, you think there are Internet police?"

"Of course not." I was annoyed because Ben was making me look like an idiot in front of my friends. "So, what's the site?"

"*Urfreetunz.* Trust me. It's easy and faster than the other downloading sites."

Trust Ben? Those two words did *not* go together. I glanced at my friends. Now Marc and Dan both looked like they were hooked by Ben's offer.

"Is it regular file sharing?" I asked.

Ben stood and came over to the computer. "It's legit, man. One of Brett's college friends created the site last year." He punched my shoulder. "Le' me sit down, moron. I'll get you started."

I shrugged and moved out of the chair.

Ben plopped down, and his fingers flew over the keys. He clicked on download program, and in less than a minute the program was on the computer and the main page was up. *Urfreetunz* scrolled across the top, its color changing every time it started across again.

"The part that takes the longest is clicking off songs you want to download," he said.

"Cool!" Marc said. He, Dan and I hovered behind Ben.

I had to admit, it *would* be nice to get all of the songs I wanted, and if it downloaded super fast, all the better.

"You see these boxes here?" he continued, pointing to a section on the right. "All you have to do is put in the first few letters of the title or artist and before you know it a list pops up. You click on what you want and, voilà," he said, spinning the office chair toward us, "it's ready to download."

"Sweet," I said under my breath. It looked almost too easy.

Ben stood. "Don't say I never did anything nice for you. I just saved you big bucks, little bro." He pretended he was going to slap each of us on the side of the head then sauntered out of the room. Dan, Marc and I turned back to stare at the computer.

"What are we waiting for?" Dan asked. "Let's put in your list."

As Ben promised, in no time at all we had downloaded almost 50 songs. The files seemed to fly into the folder on our computer. Once I transferred them to my mp3 player, I could take them off the computer. Mom and Dad would never have to know we'd used the program. The way Ben talked, it sounded like it wasn't like the older file sharing sites, anyway.

Dan's cell phone rang and he pulled it from his pocket. "Gotta go. My parents are here. Marc, they're dropping you off, too."

They got up and Marc slung his guitar case over his shoulder. "That is so cool that you got so much music, man. I'm going on that site when I get home."

"Me, too," Dan said.

I followed them to the door and yelled hello to Dan's dad. I always get compliments on how polite I am, but I only do what my parents always taught me to do. Once my friends were gone, I returned to the family room and downloaded another 30 songs. In my list journal I had put checkmarks next to each one we'd downloaded so I wouldn't do any of them twice. I still couldn't believe how quick and easy it was. It was amazing to think Ben had actually helped me for once.

Before dinner I started checking the files to make sure they all worked. I'd put them in a special folder on the computer so I'd know which ones came from that site. As I clicked through, I was surprised when a video file came up rather than a music file. I wondered if when I was selecting from the list I had accidentally clicked on a music video file without realizing it. I figured I'd check it out before deleting it in case it was something I really wanted.

My surprise turned to shock. It was a music video, all right, but it had to be rated X, or maybe even triple X if there was such a rating. I don't think I'd ever seen anything that raunchy in my life. I closed it out and clicked delete. I definitely didn't need something like that on my computer. How could something like that get on that site?

Fortunately I didn't find any more files like that. I hooked up my mp3 to transfer the music files and then left it to do my homework. It was after dinner by the time I got back to take the mp3 off the computer.

By then Mom was working on the computer. She glanced up when I walked in the room, and by the way her eyebrows were lifted and her mouth was in a skinny straight line, it was obvious something was wrong.

"Hey!" I said. "Can you click on eject for my mp3, please?"

Mom sat back in the chair and shook her head. "I'm going to start with you."

"Huh?" My heart tripped into second gear. That was definitely a 'you're in big trouble' tone.

"Pull up a chair." She pointed to the chair Dan had used earlier. I sat in it, but she shook her head. "No, bring it over here next to the computer."

Oh boy. Sweat popped out on my hands. Mom never "invited" me to sit next to her when she was working on the computer.

"Were you using the computer earlier?" she asked.

"Yeah."

"What for?"

I squirmed a little in the chair like I was a little kid. "Dan, Marc and I were downloading music for the band."

"Where did you get it from?"

"I used Aunt Meg and Uncle Donny's gift card. Why?"

"That's not the only site where you went to get music from, right?" She clicked on the icon in the program files for the *Urfreetunz* url and the home page came up. "Because there's this one, too."

I swallowed hard. Something told me I was in *really* big trouble. "Ben told me about that site." When she opened her mouth to say something, I hurried to explain. "He said one of Brett's college friends created the site and that it's legit."

She sat back in the chair and crossed her arms across her chest. "Did you pay for the music you downloaded?"

"Uh, no."

"Then it's not legit or legal, is it?"

"But Ben —"

She held up her hand to stop me. "We're talking about you. If you didn't pay for it, it's stealing, Troy. It's as simple as that."

I could feel my face turning red. Why had I listened to Ben?

"And, there's one other thing." She clicked on the trash can on the desktop and put the cursor on the video file I'd deleted earlier. "I was emptying the trash on here, and guess what I found was downloaded around the time you were downloading music."

I was afraid she was going to click on the music video file that I'd discovered with my other music files. I'd die if that started playing with my mom sitting here.

Panic clutched my stomach. “I don’t know where that came from. It was in with the downloaded music files, but I swear I didn’t do it on purpose. It was just there when I was checking them before dinner, so I deleted it.”

“Did you notice the computer was working slower?”

“No,” I said. “I went and did my homework after that.”

Mom pursed her lips then said matter-of-factly, “Well, it appears you downloaded a lot more than a few songs, Troy.”

# I.S.S.C.
# Internet Safety Savviness Challenge

Do you really understand the risks involved with file sharing? Challenge yourself to see how many of these questions you can answer.

(Answers are located at the end of the chapter.)

1. Do you know what a computer virus is?

2. How can computer viruses affect your computer?

3. What does P2P stand for and what is it?

4. If you're concerned that there's a virus on your computer, what actions should you take?

5. How does file sharing work?

6. What is spyware?

# What Can We Learn?

Troy violated one of the biggest safety rules we have; he ignored his gut feeling that doing something was wrong. When Troy felt what Ben was doing was wrong and he did not believe him, he should have stopped and looked into what Ben was saying. If Troy talked to his parents or researched for himself, he would have learned that *Urfreetunz* was a peer to peer software program and when these programs are used to download music, it is illegal. (Note: *Urfreetunz* is a fictional site.)

As a lot of kids do, Troy fell for the lure of getting his favorite music and not having to pay for it. Many kids ask, "Is it really stealing? Is it wrong?" To put this in perspective, ask yourself this: would you walk into a store and fill your pockets with CDs and leave without paying? The answer is no, because you know there are consequences and it is illegal. The property you would have put into your pockets belongs to someone else until you pay for it, and then you own it. It is the same principle with downloading music without paying for it.

Songs are copyrighted and are someone's property. When you buy a CD or pay for a download of a song, you are paying for rights to possess it. The songs belong to the artists and when you don't pay for them, you are stealing from them. This isn't always the easiest thing for kids to understand. We interpret possessing or owning something as items we can hold or physically have control over. Not everyone views songs as possessions, but they are.

Along with violating copyright laws, there are other concerns in regard to peer to peer usage. Troy discovered the hard way that not only can music be shared but so can videos and pictures. A lot of inappropriate/pornographic material is shared over these networks. Some files are clearly marked and it's obvious what they contain, but others may not be so obvious. If a person possesses inappropriate images of a child and wants to share them, they are not going to name the file "Illegal Material". They are going to be more secretive and creative. This puts innocent people who use these programs at risk.

If a child, or adult, downloads an image that is deemed to be child pornography, whether it was on purpose or not, it is a criminal act. Once the image is downloaded, it becomes *your* possession and that is a situation in which no family wants to be involved. Possession of child pornography is a very serious offense with long-term consequences and has to be taken very seriously. Even if the picture was not child but adult pornography, is this what children should be seeing? Absolutely not. Peer to peer software is not what children should be using.

If what you have learned so far has not made you re-think using peer to peer how about this: VIRUSES. It's through these networks and *free* downloads that viruses and spyware get spread like wildfire. When you use peer to peer you are allowing someone to enter your computer and take files set up in a certain folder. It does not take a lot to take a song or two and leave a virus behind. Some of the items may also have viruses embedded in them so once you download them and open the file, the virus spreads.

Another concern is that some spywares will log what keystrokes you hit and then supply them to the "bad guy". These keystroke logs can record your instant message with a relative, e-mail with your co-workers, chats you have in a chat room and also your bank account numbers and passwords you use. Here is a scenario that illustrates how spyware works.

A child gets on and downloads 20 songs, a savings of about $20.00. In the process of downloading one of these songs, spyware is downloaded that logs the keystrokes. The child logs off when all the songs are done and leaves the computer. Later, Mom sits down at the computer and logs into her bank account to pay bills. She logs in with her account number and password and pays some bills. The next day she checks her accounts and realizes that more money than she authorized is now missing. When she calls the bank they notice that a transfer of funds was made very early in the morning, and now a criminal investigation is underway. Is it worth saving $1.00 a song?

Troy was lucky. His mom had some knowledge about computers and the Internet. Catching what happened before it got out of control was a huge key in protecting her family. Troy knew better but refused to follow his gut feelings. If Troy had been patient and done some research, he could have avoided an illegal and embarrassing situation.

If you have a computer with peer to peer software, remove the software immediately. If you safely remove it you should also update your virus protection and run a full scan on your system. Hopefully you caught it in time before viruses have been installed.

A good way to avoid these situations is by remembering this famous old quote: "If it sounds too good to be true…it probably is."

# ISSC Answer Key

1. A computer virus is a program that is loaded onto your computer, usually without your knowledge, that attempts to use your computer in ways that you most likely don't want. Some viruses will affect your ability to use your computer properly.

2. There are several ways viruses can affect your computer. One way is they can slow down your programs. If your computer won't start up Windows™ then it's likely it's been infected with a virus. If the Internet is working slowly or not at all, again, a virus could be the culprit. One major indicator of infection is if your desktop background or homepage have been changed and you didn't do it. (These symptoms don't always indicate your computer has been infected with a virus, but it's recommended that you check it for viruses.)

3. P2P stands for Peer to Peer networking and it means that there is a connection or sharing of files between two direct computers instead of using a website as a means of connection. Common examples are Limewire, Bearshare, Kazaa and Frostwire.

4. If you are able, back up any of your important information

(documents, pictures, computerized financial records, etc.) This will ensure that if the virus causes the Operating System to have to be reloaded, you will not lose your information or have to pay a repair company to back up that information for you. Once all of your information is backed up, try using your anti-virus program to remove the virus. If this fails, you will need to visit a computer repair company to attempt to have the virus removed.

5. When file sharing over the Internet, you are opening up your computer to strangers. When you connect to these file sharing programs, you are literally granting access to your files without requiring a password or any other way of verifying who is viewing your information. When you and another file sharing user have the

program open on your computers at the same time, it's like opening a door and saying to them, "Take whatever you'd like."

6. Spyware is much like a virus, except that it rarely affects your ability to usc your computer. Instead, it tracks everything that you do on your computer (websites visited, passwords entered, credit card information, etc.) so that it can either steal your information or allow companies to directly target you based on your interests and spending habits. Spyware can be found on even the most trusted websites, so it is recommended that you maintain and update anti-virus and spyware protection on a regular basis.

# CHAPTER 4

## Patti's Pandemonium

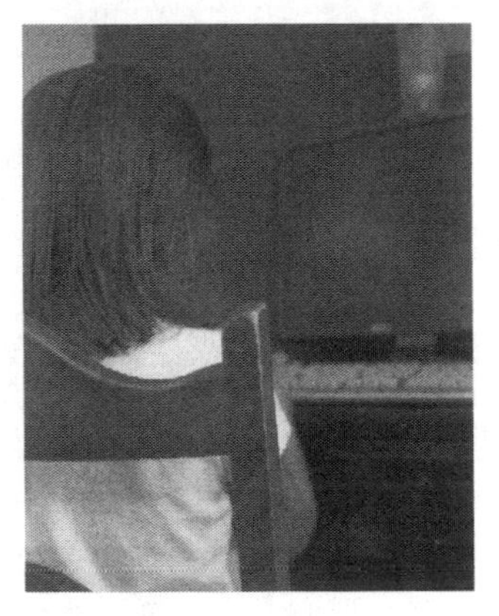

**Profile:**
8th grader
Class president
Mother: Freelance Journalist
Father: Construction Worker
Siblings: one older brother, two younger sisters
Town demographics: medium-sized town, suburb of a city

I can pinpoint the exact day that a decision I made changed the rest of my life and the lives of three of my friends. I'd always been a good decision-maker, and in fact, I think that's one of the reasons I was re-elected class president. I have good ideas and I'm pretty good at getting others to stand behind me and help turn those ideas into action. So even though in reality I can't take responsibility for the choices of others, I'll now admit none of this would have ever happened if my friends weren't so loyal.

On that day, as usual, the school halls were crammed with kids changing classes when I stopped by my locker to get my math book. Also as usual, Beth, one of my best friends, was waiting for me, which was perfect because I had a "job" for her. She was standing in front of my locker with her back to me as she talked with Shannon. I hip-bumped her out of the way.

She stumbled a bit as she turned toward me. "Watch it!" she snarled before she realized it was me. Then her tone changed. "Oh, hey!"

I spun the dial to put in the combination on my locker. "I need you to do me a fave."

"What this time?"

"Hey! Is this my bff sounding cranky?" I looked past her to Shannon. "I'm sure Shannon would be happy to help me out." Shannon wasn't really one of my friends but she hung around Beth between classes lots of times, so I put up with her.

Beth laughed. "No, it just seems like I'm always doing *something* for you."

"That's what friends are for, right?" I gave her my best "chum" smile.

Beth shrugged. "Whatever. So, what's the favor?"

I didn't really want anyone else to know what I was having Beth do. The fewer witnesses, the better. I tipped my head toward Shannon and whispered, "Has to be just between you and me, if you know what I mean."

"Oh, right." Beth turned toward Shannon. "Hey, Shan, I'm not feeling so great, so I'm going to stop by the nurse's office on my way to Spanish. Can you tell Señora Chapman I might be a minute or two late?"

If Shannon knew Beth was trying to get rid of her, she didn't let on. "Sure, no prob," Shannon said, then turned and weaved through the crowd in the hall.

I glanced around to make sure no one else was watching then pulled a note out of the front zipper pocket of my binder. It was folded over until it was a neat little square with a pocket to seal it. "Put this in jerky Jocelyn's locker."

Beth's eyes twinkled. "Sending her another message?"

My locker clanked open and I reached up to take my math book off the top shelf. "She obviously didn't get it the first time when I told her Derek was off limits."

Beth laughed. "When *you* told her? I was your messenger that time, too, Patti."

I put on my best pouty, *'Oh, I'm so hurt'* face then said, "Is that a problem?"

"No, not really, I guess."

"Then just do it. I want her to have it before lunch." I slammed my locker shut and twirled the combination lock. "She's locker 1029. No lock on it yet, so you can just open it and leave her my gift."

Just then Derek came around the corner, and I saw him past Beth's shoulder. I smiled and gave him a little wave to let him know I'd spotted him. Although we weren't officially "going out" yet, he had been paying a lot of attention to me, so I thought he might ask me any day.

"Here comes Derek. Does my hair look all right?"

Beth laughed. "When doesn't your hair look all right?"

Derek stepped past Beth and next to me. "Sorry I'm a little late, Patti. Jocelyn needed help with her locker."

I shot Beth a look that said *Jocelyn is done.* I couldn't believe this girl moved into the school less than a month ago and already she was nothing but problems for me. Every time I turned around Jocelyn was hanging around Derek. I hoped that note would be enough to get her to back off.

"See you after school," I called over my shoulder to Beth as Derek and I started down the hall. I was tempted to say something to him about Jocelyn, but I decided my friends and I would handle the problem with her if she didn't clue in soon.

At the end of the day, the class officers for seventh and eighth grade had a brief meeting with Mrs. Becker, the principal. I was psyched when it was over because Derek was waiting outside the office for me. We walked to my locker before he had to go to Run Club.

"What time does Run Club get out?" I asked.

"Four-fifteen. Here, let me get those for you." He took the stack of books from my arms and carried them. It was the sweetest thing anyone had ever done for me.

"Do you want to meet at The Shake Shack for ice cream afterward?" I asked.

"Sure, for a little while. I have a lot of homework."

We stopped in front of my locker, and I took my books from him.

"Awesome. I'll be there at 4:30."

He nodded.

"And thanks for waiting for me and carrying my books."

"No prob."

I watched him walk away, thinking how lucky I was. The cutest boy in eighth grade was hanging out with me. He was almost to the end of the hall when Jocelyn appeared from the other direction. My muscles tensed as I watched her reach out and touch Derek's arm and then everything around me went red. Either she hadn't gotten my note or she was totally dense! As much as I wanted to yell down the hall to stop her and Derek from walking away together, I was unable to form any words.

After dialing in my combination, I yanked open my locker. A whole bunch of pieces of paper fluttered to the ground like confetti. I glanced around, wondering if someone was pranking on me. I picked up a couple of the bigger pieces and stifled a scream of fury. There was no mistaking what this was. I recognized the little bits of my handwriting on the shredded pieces. My frustration was building like a volcano, and I was about ready to blow.

I felt around in the side pocket of my backpack until I found my cell phone and pulled it out. My hands shook as I tapped on *new text message* and checked off three of my friends, Beth, Maria and Sara, as recipients.

***meet me out front under tree taking jocelyn down***

I pressed "send" then glared down the hall toward the area where Jocelyn had met with Derek. She'd be sorry she messed with me. A plan was already forming in my head. Next I sent Derek a text.

***karen rogers needs jocelyn's cell number u have it?***

I'd heard Karen Rogers was in Jocelyn's lab group for a science project, so I hoped Derek wouldn't question me about her wanting the number. I figured he probably hadn't made it to the gym yet and if Jocelyn was still with him he could ask her for her number. I hoped he didn't know it himself or I'd freak, but how would I know? There was an immediate response.

***413 555 8789***

I shoved my phone in my pocket and turned back to my locker. Mr. Dietrick, my art teacher, came out of his classroom just as I finished putting my books in my backpack.

"Hey, Patti," he said as he approached. "Congratulations on winning class president again. You're a real go-getter, so I'm sure you'll continue to represent the class well."

"Uh, thanks." His compliment felt awkward considering what I had going on in my head. He continued down the hall, and I hurried out of the building. As I came down the front steps I glanced across the street toward the baseball field. On the far side of the field I could just barely see the dugouts for the teams. When it was off-season, the two dugouts were pushed together. From a distance, the two together looked like a little house, except there was a tall opening in the middle where the two sides didn't quite meet. That would be the perfect location for carrying out my plan.

Beth, Maria and Sara were chatting by the tree when I came out. Sara saw me first and came toward me.

"Patti, did you see Jocelyn was with Derek after school?"

"Yup." It annoyed me to think others had seen her with him, too. "That's okay. I've figured out how to send my message loud and clear. When I get through with her she'll know who has dibs on Derek." I hooked my backpack over my shoulder. "Come on, girls. We're going shopping."

They looked at me with confused expressions. "Just trust me. When we get through, Jocelyn's going to feel like —" I laughed, not bothering to finish my thought.

On the way to the grocery store I explained my plan. The girls were ready to help me teach Jocelyn a lesson. When we finished with her she probably wouldn't want to show up at school again, let alone try to get Derek's attention.

All I needed for my plan was a few containers of chocolate frosting, shaving cream, and eggs. I made my purchases and we left the store.

"Tomorrow before school meet me in the dugouts on the baseball field. Keep out of sight so Jocelyn doesn't see you're in there. Maria, you bring the mixture of frosting, shaving cream, and eggs with you. I'll make sure Jocelyn shows up."

Sara looked at me skeptically. "How are you going to get her there?"

I smiled coyly. "That's a piece of cake."

They walked back toward the school, and I walked in the other direction toward The Shake Shack.

I was waiting in a booth when Derek got out of Run Club. He slid into the seat across from me. I was nervous, wondering if I could think of enough things to talk about.

"How was Run Club?" I asked.

"Good. We have a cross country meet next Saturday." He picked up one of the plastic menu cards and started looking at it.

"I bet you'll be great."

He glanced over the menu. "Thanks." His cheeks looked a little pinker like maybe I'd embarrassed him.

"Yeah, well —" I picked up my phone and opened the text Derek had sent me with Jocelyn's phone number and kept it on that. "Aw, my phone died," I said, laying it on the vinyl seat next to me. "Can I borrow yours to send my parents a text to let them know where I am?"

"Sure."

He handed me his phone. I called up a new text message then glanced down at my phone's screen and punched Jocelyn's number into Derek's phone and left a message.

***meet me in dugout 2moro b4 school***

I smiled and started to hand the phone back to him when it beeped with a return message. I quickly pulled it back.

"Oh, probably my parents." I clicked on the message and saw the simple single letter response from Jocelyn.

***k***

I deleted it then handed the phone back to Derek. "Yup, just my parents. Thanks."

I couldn't believe how easy this plan was turning out to be. Derek had no idea he'd just helped me set up Jocelyn.

The next morning Beth, Maria, Sara and I met at the dugouts early. I took the mixture of chocolate frosting, shaving cream, and eggs that Maria had put in a plastic container and transferred it to a doggy poop bag. Perfect! I pulled on a couple of the plastic gloves my mother used for cleaning, then we waited.

I peered through the opening between the two dugouts and saw Jocelyn coming across the baseball field. She looked really happy like this was going to be the best morning of her life. Boy, was she in for a surprise!

"Get out of sight and be quiet," I whispered to the other girls. We crowded up on the benches and into the corners so Jocelyn couldn't see us until she was inside. Sara and Maria giggled, but one shushing look from me and they covered their mouths and held it in. The muscles in my neck hurt from the tension. It was like there was a spring inside me coiled up tight and ready to pop. I wished she would walk faster.

Finally I heard her footsteps outside. "Derek?" she called. She waited a few seconds then said his name again. We remained silent, and luckily Sara and Maria managed to not burst out laughing. Jocelyn squeezed through the opening between the two dugouts. "Derek, are you in here?" As soon as she was inside, we leaped down from the benches.

She whirled around and her eyes were really opened wide. "Wh-what's going on?"

I handed Sara my cell phone. "Record it!" I shouted as Maria and Beth grabbed Jocelyn's arms and pushed her onto a bench. Out of the corner of my eye I saw Sara flip my phone open and hold it up.

Jocelyn kicked and screamed, trying to get away from Maria and Beth. I grabbed the plastic doggy bag with our gooey mixture and held it up close to Jocelyn's face. I stuck one finger in the bag and scooped out a bit of the mess.

"This is what Derek thinks you are."

"What is that?" she screeched, struggling even more against my friends' stronghold.

"What does it look like, Jossie?" I taunted. I scooped more out with my gloved hand and slowly started toward her face. She squirmed, trying to get away from me, but Maria and Beth held her tight against the wall. Her foot flew up and hit me in the side. The air in my lungs went out on a whoosh and I fell into her. Jocelyn's head cracked against the wooden wall of the dugout.

All I could hear was her screaming and the other girls laughing. I picked myself back up and smeared the mixture all over Jocelyn's face.

"No! Stop!" she screamed. "Stop. Please."

I knew no one would be able to hear us out here, so I didn't care how much noise she made. "I told you to stay away from Derek." It was hard to catch my breath as I moved around to avoid her kicking feet. "You go near him again and this video is going to be seen by a lot of people."

Maria, Beth and Sara laughed hysterically as I wiped all of the disgusting concoction on any of Jocelyn's exposed skin. At one point I saw Sara lower the camera phone.

"Don't stop recording!" I bellowed. She raised the camera to capture the last few seconds of my masterpiece. When I was done I stepped back to admire my work. If Jocelyn didn't get this hint, then she was totally clueless.

"Let her go," I ordered my friends.

As soon as they released her arms and legs, Jocelyn curled up in a ball on the bench and sobbed. For a few seconds I felt bad for her, but then I thought of Derek. I'd tried to warn her gently with the notes. She didn't get it. We had no choice but to send a stronger message.

"Come on, let's go." We left Jocelyn whimpering on the bench and climbed out of the dugout. Once outside, I turned to see Sara punching numbers into my phone.

"What are you doing?" I asked. I tried to snatch it from her hands but she turned away.

"I'm sending it to our phones, too," she explained. "I want a copy. Everyone's going to love this."

"What? No!" I grabbed at it again but she had hit send before I got it from her.

Buses were starting to arrive. We ran across the field. When we got closer we slowed down and walked so we wouldn't look suspicious. Every now and then I turned around to see if Jocelyn had come out of the dugout, but there was no sign of her. Maybe she *had* learned her lesson.

For the rest of the day I watched for Jocelyn, but she never showed up at school. Throughout the day a few friends, and people who weren't my friends, made random comments or gave me strange looks that made me wonder if they'd seen the video, but how could they?

At lunch time I grabbed my cell phone from the pocket of my jacket and checked my sent messages. I wanted to scream when I saw that Sara had hit my "friends list" as the recipients for the video text from my phone. That meant dozens of people had received that first thing this morning. I wondered if Derek had seen it too.

I found Sara at our usual lunch table. It was the first I'd seen her since before school. I dropped on the bench next to her and leaned in close.

"What were you thinking?" I tried to keep my voice low enough so she would know how angry I was but no one else would hear me.

"What?"

"You sent that video to everyone on my friend list," I hissed. "Do you have any idea how many people may have already seen that?"

Sara swallowed the food in her mouth then shrugged. "What was the point of recording it if you didn't want anyone to see it?"

"It was just for me to use against her, you idiot! No one else was supposed to see it."

"I thought I sent it to me, Maria and Beth," Sara said.

I shook my head and got up from the table. "No, you sent it to everyone in my phone book on my cell. Thanks a lot."

Instead of eating lunch in the cafeteria I ate outside. I took my phone out with me and sent a text to everyone on my friends list.

***Don't do n e thing w/ video txt. Plz delete***

I hoped my friends all respected me enough to do as I asked. I could only use the video against Jocelyn if I was the only one with it.

That afternoon when I got home the first thing I did was go on-line onto my *FriendPost* site to see if there were any comments about her. I checked the news feed and right away saw that Nolan Warwick, a kid I used to like, had posted a video under his profile. The message he'd put under it was:

***someone is in deep doo doo - lol***

I clicked on the video and saw there were already hundreds of views. I recognized Jocelyn's screaming and the clear video from my cell phone. I watched the whole thing in shock. I couldn't believe someone had posted the video already and that so many people had seen it.

A lot of the comments under the video were funny ones that made reference to what the brown mixture looked like. Some people didn't think it was as funny as we did, though, and left comments saying they thought what we did was wrong or stupid. They could think what they wanted. I got my point across, and that's all I cared about. Despite the few negative comments, I decided maybe it was okay that this was posted, because now she would be too embarrassed to see Derek.

The next day I realized I should have cared more about how many people were seeing the video. I'd only been in homeroom for a couple of minutes when I was called to the office. As soon as I walked in I was escorted back to Mrs. Becker's office.

My heart jumped to my throat when I walked in. My mom sat in a chair across the desk from Mrs. Becker. To the right by the windows were two police officers. One was Officer Nash, our school resource officer, and the other was a female officer I'd never seen. I looked back and forth between them, my mom and Mrs. Becker.

"Mom, what are you doing here?"

Mrs. Becker leaned forward and pointed to the empty chair next to my mom. "Have a seat, Patti."

The hair on my arms stood on end. I looked from Mom to Mrs. Becker to the police officers.

"Wha-what's going on?"

The female officer propped her hand on her hip and said, "That's what we want to know, Patti."

# ISSC
# Internet Safety Savviness Challenge

In the scenarios below, put yourself in the situations then circle Yes, No, Never, Maybe or I Don't Know in response to how each of these relate to you.

**Have you ever…**

1. …sent a text, instant message or e-mail to someone with content that you later regretted?

Yes No Never Maybe I Don't Know

2. …forwarded a picture or video of someone that you thought was funny but you didn't have permission to forward?

Yes No Never Maybe I Don't Know

3. …received or been the subject of a text, e-mail, picture or video posting that made you feel uncomfortable or bad?

Yes No Never Maybe I Don't Know

4. … seen a picture, video or comment that was hurtful to another and reported it?

Yes No Never Maybe I Don't Know

5. …forwarded a picture, video, comment or e-mail that made someone else feel bad?

Yes No Never Maybe I Don't Know

6. Do you know what the legal consequences are for cyber bullying?

Yes No Never Maybe I Don't Know

7. In the future if you witnessed a situation similar to Patti's and Jocelyn's would you ask an adult for help?

Yes No Never Maybe I Don't Know

# What Can We Learn?

Patti demonstrated a behavior that has been around for years and unfortunately probably will never go away - bullying. Parents can remember what bullying was like when they were kids. Most times the bully was the biggest boy in the grade or school. He would get his way because he was bigger and used intimidation and fear to get what he wanted. Now, with technology and the Internet, there is no longer a stereotypical bully.

Bullying comes in many formats and styles but one thing will always be consistent: it should never happen. Even in the past, before the Internet, when someone was being bullied it was not a fun situation at all. Back then kids might have a behavioral routine if they were victims. If someone was being bullied, he or she might fake being sick the next day so he would not have to go to school. He would stay home where the bully could not get to him or bother him. Usually after the bully's target missed a day of school the bully would be on to someone else. This is a cycle that no kid should have to live through, but with today's technology it is worse and even easier to bully.

If a child gets bullied at school and goes home, to a place where we should all feel safe, they might not be able to avoid the situation anyway. Through e-mails, chat rooms, picture hosting sites, text messaging and social networking sites, a child can't escape.

In this case, Patti got home and realized how quickly this situation had spread; think about how fast Jocelyn heard about the video being posted. Once this video was forwarded, no one had control over where it went or who saw it. Jocelyn did not go to school after this terrifying incident. Who would? She probably went home, upset, sad and frightened. When she got home she probably received texts from a lot of people. Some of the people would be making fun of her and making comments they would never say to her face but over texting they would. These comments would make Jocelyn feel worse, if that was possible.

Undoubtedly, some text messages would also be supportive, coming from friends who are asking Jocelyn if she is okay. We all know this is how a true friend would react, but put yourself in Jocelyn's shoes. She is humiliated, sad and hurt, and now she knows that a lot of people have seen what happened. As well-intentioned as they are, she may not want to hear all of these comments right now. In these situations, if kids want to be supportive they should get parents involved as soon as possible. This is the best help a friend can offer.

Receiving the text messages alone would be difficult for a victim to deal with, but when we add the social networking and Internet it becomes a much bigger problem. Videos on the Internet almost never disappear permanently. When Patti found Nolan's site and discovered the video was on there, she should have realized right then what a terrible mistake she had made. If Nolan had it, numerous people saw it on his site and downloaded it to their computer which shows how things on the Internet never go away. Someone, somewhere, will possess this video for years to come and it will re-surface from time to time.

On Nolan's site people were making funny comments about the situation, comments most would not say to Jocelyn's face, or in person to anyone for that matter. Every day people say things on the Internet that they would NEVER say in person. Why is that? Do people really think no one will know who said it?

When pictures, videos or comments are posted on the Internet, sometimes they may appear anonymous. It's common practice for Internet users to use screen names instead of real ones, post cartoon pictures for a profile instead of using a real picture or even create entirely "fake" social networking sites in order to avoid revealing their identity. When people engage in this kind of behavior they think they can never get caught. Maybe at first parents, law enforcement or school officials aren't able to identify the person, but other kids can, and will. Sooner or later the adults will know who made the comments or posted a picture. If a law has been broken because of harassment, threatening or other crimes, law enforcement can apply for court orders and they *will* ultimately identify the person.

To avoid this situation, there is a very simple rule to follow if you're considering saying something about someone: if you would not say it to the person's face or say it in front of your parents or teachers, you should not say it. PERIOD.

In the scenario presented, the school administrators took a crucial course of action by getting law enforcement and Patti's mother involved. While the school will no doubt mete out their own punishment, Patti will probably also face serious criminal charges because laws were broken. The charges could include harassment, assault, intimidation, disturbing the peace and threatening. (Each state has specific laws for these types of situations and will charge accordingly.) In the end, Patti will most likely have to make appearances in court and perform community services. Besides the legal factor, Jocelyn was hurt in a way no one should be; Patti needs to learn from this.

This type of incident affects many people. For example, another aspect to consider is how Patti's parents will feel when they see Jocelyn's parents around town. All of these consequences could have been avoided if Patti had just thought ahead or if Sara or Maria had done the right thing and told Patti she was wrong.

And that brings us to Sara, Beth and Maria. They are just as guilty as Patti. When bullying occurs and others get involved instead of putting a stop to the behavior, they have a level of responsibility, too. In this case, Sara, Beth and Maria were accomplices and need to be punished and most likely will be. But their actions may also have far-reaching effects on others close to them. For example, what if one of these girls has a younger sister who comes to the school the next year? Don't you think people will talk and wonder if she is as mean as her sister? Or Patti, Maria and Sara may be on a school sports team and when this incident is over, they may be removed from that team. The consequences for their behavior could affect them in more ways than they can imagine.

As the victim, Jocelyn will also have a long journey ahead of her as she tries to put this behind her. She is only in $8^{th}$ grade now but this incident will follow her for a long time. Parents, can you think back to when you were in elementary school and incidents happened that you still remember? Kids fighting? A fellow student bullying another? A teacher yelling at someone a certain way? I am sure we can all recall a bullying situation, whether it was a major incident or something small. Probably the incident you recall is one that you wish you never saw or were involved in. These recollections are just from our memory. Our children's generation will have pictures and videos on the Internet that will document the things they want to forget. It is important that we take an aggressive approach to stop cyber bullying.

If anyone, parent or child, knows of a situation where someone is being bullied, cyber or not, you need to get involved. Kids, you MUST let your parents, teachers, principals or trusted adults know if you are aware of someone who needs help. Although it showed their concern for her, Jocelyn needed more than just friends asking if she was okay. She will be dealing with emotions and feelings that friends may not be able to help with, but adults may. Parents, if a child tells you she thinks someone is in trouble, especially if the Internet is involved, this can't be ignored. The quicker you get involved, the better. If you don't know what to do, contact the school or law enforcement authorities and let them determine the extent of the concern.

If parents, schools and law enforcement work together it accomplishes a great deal. First, it sends a message to the victim that people care about her and will work together to help her. Second, it sends a strong message to the community that this type of behavior is not going to be tolerated. If actions are taken quickly it may slow down the circulation on the Internet. If kids hear about consequences and realize that everyone is taking the situation seriously they may not make negative comments or download a video that they know may get them in trouble.

These situations can have long term negative effects on everyone involved. It is very possible that when Jocelyn goes to college she will be in a lecture hall with students from all over the country, including her home town. While listening to her professor and trying her best, she could hear from behind her, "That's the girl from the deep doo doo video." If you were Jocelyn, how would that make you feel?

There is no place in anyone's life for cyber bullying. The well-known African proverb, "It takes a village to raise a child" can apply to cyber bullying situations as well. If someone is being bullied in any way, it is the responsibility of anyone who knows about it to take action to stop it. The "www" on the Internet stands for World Wide Web, and we all navigate in the villages that make up that World Wide Web. We need to do our part to make sure that no child in the village becomes a victim.

# CHAPTER 5

## David's Difficulty

**Profile:**
14 year old eighth grader
Loves playing sports
parents divorced/ custody split
Mother: Paralegal
Father: Surveyor
sister: Elizabeth, 16
Town demographics: 1, 300 residents, rural community

As my dad says, I've learned a lot of lessons the hard way. For example, one lesson I learned the hard way was to always wear a helmet when skateboarding. I learned it the *hard* way, because I found out concrete sidewalks aren't forgiving if your flip kick doesn't work out and you miss the board. When you lose your balance, fall and the back of your head smacks the cement, it's pretty hard. A concussion and eighteen stitches later, I learned to wear my helmet, especially if I was going to try tricks. I still have a two inch scar where the hair never grows in right. That was a lesson I learned the hard way – literally.

There have been lots of others, too, but none of them were as bad as the latest one. Like the scar on the back of my head, my stupid mistake on the Internet was something I learned from the hard way. The scar I have from the Internet incident isn't something anyone can see, but I know it's there every time I think about posting a picture.

Every story has to start somewhere, and my story started on that fabulous snow day in March. The early spring snow storm caught everyone by surprise. I heard the great news when my radio alarm went off. The D.J. was in the middle of the list for school closings. He'd already passed the Bs and was on Gilbert, a town about twenty miles away, so I knew it would be a while before he got back around to Branford. A snowplow roared down the street confirming there was a good chance my school was cancelled, too.

I glanced toward my desk at my computer. In less than ten seconds I could check my school's website and know if we were one of the lucky ones. I kicked back the covers and hopped out of bed. My computer had gone to sleep, but with one swipe of the mouse, the dark screen filled with color. I dropped into the black vinyl swivel chair and typed in the school web address. The banner I'd hoped for was there: *Branford Public Schools Closed.* I was going to go and wake up Elizabeth, my older sister, but she can be crabby in the morning, so I decided to let her figure it out herself – or let Dad risk *his* life opening her door.

My gym bag holding my baseball cleats, my glove and practice clothes was sitting next to my bedroom door right where I'd left it the night before. I was supposed to start baseball tryouts for the middle school baseball team that day. Although I was bummed that practices would have to wait until the snow cleared off the field, it was pretty cool to have an unexpected day off from school. That also meant I'd have an extra day to cram for the test in Mr. Cook's science class, although in reality, I knew I probably wouldn't spend more than five minutes going over my notes again.

I flopped back onto my bed thinking maybe I'd go back to sleep, but I was too wired knowing I now had a whole free day to do whatever I wanted. Rolling onto my side, I noticed a slice of sunlight came past the side of my window shade and was sparkling off the gold basketball player on my trophy. Recently, my travel team had won the big tournament at the end of the season and each of us received our own trophy. That had been an awesome weekend I'd never forget. In two of the games I'd had a record number of steals, including three really good passes that I'd picked off from the star of the other team. It got to him so much that he wouldn't even slap my hand when we went through the line at the end of the game.

Suddenly I remembered all the pictures Dad had taken that weekend. I'd promised the guys on my team that I'd get them up on *Pics4Sharin.com,* a photo sharing site. I hadn't been at Dad's house since then. Dad had promised me he'd load them onto his computer before I came back. Since our computers here were networked, I could access them from the computer in my room.

Climbing out of bed, I went back to my desk. I supposed getting those pictures up was one thing I could accomplish thanks to the snow day. With just a few clicks I was into the special family pictures file Dad kept. He was an organized guy, and as I expected, he had the pictures filed under *Pictures – David – Hawk's Basketball Tournament.* There had to have been over a hundred photos in that file.

I started clicking through them. Most were sick action shots of me dribbling and shooting, but there were also good ones of Andy and the rest of my teammates. Most were taken at the basketball court, but then I laughed out loud when I got to the ones taken at the hotel after the game. Since Andy's parents couldn't go to that tournament, Andy went with my dad and me. We paid for a cot for the hotel room so he'd have his own bed, too.

After the game we'd gone back to the hotel and the whole team was going to meet at the pool and have pizza and swim. Andy and I changed into our trunks and were waiting for Dad to order the pizzas.

While waiting, we goofed around. There was a huge mirror on the sliding closet door. I wasn't wearing a shirt, so I stepped in front of it and started posing, flexing my muscles and trying to look really buff. For the last few months I'd been working out using my dad's gym equipment and going to the weight room at school, but I hadn't exactly gotten the results I'd hoped for, yet. Even though some of the other guys on our team and other teams looked like they were the sons of The Incredible Hulk, I figured if I kept working out, eventually I'd look buff, too.

When I wasn't looking, Andy grabbed my dad's camera and started taking pictures of me. We were laughing so hard our guts ached. Now, as I clicked through the pictures, I could see my face was getting redder from laughing so much. I hadn't laughed like that in a long time. There were also pictures Dad had taken of all of us in the pool. Just looking at the pictures got me pumped again, just like I felt the night we won.

I finished going through the slide show then went back to decide which pictures I was going to post to the *Pics4Sharin* site. I made sure I had pictures that included all of the team. I even made a list of each of my teammates then checked them off once I'd found at least one good picture of each of them to put up.

Like a lot of my friends, I'd started my own *Pics4Sharin* account at the beginning of eighth grade. I opened my account and clicked on Hawk's Basketball. It was a group site where anyone could upload pictures. There were already some pictures there, so I made sure the ones I uploaded were different. I finished clicking on the ones from our file that I was going to share when I came to the "muscleman" pictures again. Yeah, I was kind of skinny and scrawny but the pictures were hilarious, so I clicked on the one where I was posing like those old Egyptian or Cleopatra pictures. The basketball was balanced on my arm like it was my muscle and my teeth were clenched like I was growling. I was sure it would give the other guys a good laugh, too.

I signed on to my *FriendPost* account to update my status with the message:

***hoops pics up on Pics 4Sharin check it out***

I also posted the link under my status so I wouldn't have to tell all of my friends individually. On the way out of my room I opened the window shade. The sun glared off the basketball trophy, blinding me. It should have been a hint about the kind of spotlight I would be in a few days later.

It was my sister, Elizabeth, who gave me the heads up. Her boyfriend, who is my coach's son, sent her a link to a site he'd discovered. She sent the link to me with a warning that I wasn't going to like it. I plugged it into the browser.

*Hawks' Secret Weapon* came up first as the file title, then the pictures loaded. I swallowed hard against the lump forming in my throat as one horrible picture after another came up. I recognized the pictures as being some of the ones I had put up on *Pics4Sharin*, but they didn't look the same as when I'd uploaded them. By the time the file finished, there were eleven pictures of me that had been photoshopped. My attention was glued to the screen, shocked that someone would do that to me.

In the picture of me where I was holding the ball as my muscle, the person had made a tattoo on my chest that said "I like men". I remembered how hard Andy and I had laughed that night when he took that picture. I wasn't laughing now. In the other pictures, who ever had done this had changed my uniform to have rainbows on it. One showed a picture of me with lip prints on my cheek and Adam Lambert photo shopped in making eyes at me. In another one it looked like I was pinching another player's butt. The pictures got worse as I clicked through. I groaned and dropped my head against the back of the chair. Anybody who didn't know me would think I was gay.

I then realized that people were posting comments under some of the pictures. I went back to the first picture of me with the tattoo. Under this picture alone were 15 messages. Oh, no! How many people had seen this already? Even though I didn't want to read what was being said, I had to know.

***Trevor211112 (2:05 pm): Did any boys play in this tourney? LOL***

*SaraLikeU (2:03 pm): LOL WOW and he wuz on winning team, wunder what losing team looked like. ROFL*

*LUVHrses21 (2:02 pm): This is NOT COOL, take these pics down, David is nt like this*

*Trevor211112 (2:00 pm): Wondr if his team let him shower with them LOL*

*SaraLikeU (1:55 pm): LOL how com he is not 1 of the cheerleaders LOL*

*BBallGuy (1:53 pm): We LOST TO THIS DUDE? If I knew I wuld ve brought him flowers LOL*

*MadMatt54678 (1:45 pm): U guys suk. David is very cool, Take these Pics DOWN!*

*Rockn09876 (1:22 pm): This is y I dont play sports LOL get a life*

*SaraLikeU (1:17 pm): Bet his mom helpd him pick out tattoo LOL ☺*

*Sk8rgrl64322 (1:06 pm): Come on guys knck it off. Take this down*

*WldnCrZy1 (12:55 pm): Dude is HOT!!!!! for a weirdo LOL did ur bf take these pics. LOL smile 4 camera FREAK!!!!!!*

*CasseChilln (12:50 pm): WOW I nevr knew this bout u. no wondr u dont hve a gf*

*Prty4u0987234 (12:25): Wht a loser!!!!!!!!! Does ur bf know bout the tattoo LOL*

*WldnCrZy1 (12:16 pm): Hey look, it is our school cheerleader captain. Quit the man sports dude, hey I thnk thy hve a sewing club at school, u culd b prezident LOSER*

***BranfordRocks2009 (12:03): WOW, dude???????***

My stomach flipped, and I thought I'd throw up. I've never done anything like this to anyone, so I couldn't understand why they would do this to me. I get along with everyone; well at least I thought I did. This was bad, but then I saw the worst comment of them all.

Under the picture they changed so it looked like I was pinching another player's butt, there was one comment. I knew the screen name. It was Trevor. He was a junior and the captain of the Branford High varsity basketball team. They had won the state championship last year.

I met Trevor at a basketball camp this past summer and we got along well. We talked about how cool it would be to play together next year at the high school, him as a 12th grader and me as a 9th grader. He even told me that he thought I would make the team without any problem. Ever since I started playing basketball I have wanted to play on the high school team. When I saw his comment, I couldn't believe what he'd written.

***TrevBan2010 (11:30 am): What?????? No way we want him on team. We dnt need some guy grabbin us in the shower. Met him at camp, hes not that good anywy. Gona show coach these, maybe he could be a cheerleader 4 us instead. Looks like he can dance and hang with the girlz.***

I was shocked. Even though it made me feel like a baby, I stared at the screen and couldn't keep from crying. I hadn't done anything wrong except have fun with a friend and post what I thought was a funny picture. There was no way I could go to school tomorrow. No way!

I wondered how many of the guys on my team, or any of the teams we play, had seen these pictures. It didn't take long for me to get my answer. When I checked my inbox, I had a ton of e-mails from friends and people I didn't even know. The first e-mail I received was from a good friend, Maria.

***To: David (BBall4life65321)***
***From: Maria (ILUVHrses21)***
***Subject: u ok?***
***Hey, nt sure if u know this yet but there are some bad pics of u up on pics4sharin. People r talkin bout em. U r gona wanna check em out. Call me l8r***
***Xoxo***
***Maria***

I wanted to die. Maria's been one of my good friends for a long time. I know she would never do this to me, but the fact that she had seen them was humiliating. And, if she saw them, everybody must be seeing them.

I looked at the next e-mail. The subject said, *"WOW fairies playBBall?"* My heart sank, and I didn't bother to open the message. I closed down my e-mail, slumped in my chair and felt totally helpless.

# I.S.S.C.
# Internet Safety Savviness Challenge

Test your Internet safety savviness and compare scores with your friends or parents.

**1. Review the comments under David's pictures on the Internet. Based on what people said, who has David's best interests in mind? (Circle all that apply.)**
**a)** Andy **b)** Trevor **c)** sk8trgrl74322 **d)** Maria **e)** none of them

**2. Whose actions helped David? (Explain your choice.)**
**a)** Andy **b)** Trevor **c)** Dad **d)** Maria **e)** none of them

**3. Who has access to David's pictures? (Circle all that apply)**
**a)** his teammates **b)** his friends **c)** his dad **d)** Trevor's H.S. teammates **e)** anyone with Internet

**4. How long will David's pictures be available on the Internet?**
**a)** indefinitely **b)** one month **c)** six months **d)** one year **e)** one week

**5. Who should David talk to for some help in this matter (Circle all that apply)**
a) police b) teacher c) principal d) parent e) friend

**6. What steps should David take to preventing this from happening again?**
a) be careful what camera he uses when he takes pictures
b) make sure that no one gets his computer password
c) never post pictures to a public style web site
d) never trust anyone ever again

**7. If David went to his parents and told them about this situation, what are the best decisions for the parents to make in order to help? (Circle all that apply)**

a) Take the computer away from David and never allow him to use the Internet again.

b) Contact the school system and advise them of the situation

c) Contact the local police department and report this incident

d) Contact all of David's online friends and threaten to call the police

e) Send e-mails to everyone who posted comments telling them how wrong they were

**8. If one of David's friends really wants to help, they can (Circle all that apply)**

a) tell their parents that David may need help.

b) tell a teacher or trusted adult that David is in trouble.

c) talk to David and show him support.

d) talk to David's parents and advise them of the situation.

# What Can We Learn?

***I glanced toward my desk at my computer***

Parents, it is essential to keep the computer in a public area of the home. In this situation, it would have been very helpful, on many levels, for David to have a parent by his side. He is dealing with a lot of emotions after seeing what people said about him. Being alone in his room, scared, upset and confused is not going to help him or resolve the issue. Situations on the Internet happen very fast, and that's why parents should be around and available to help.

***I finished going through the slide show then went back to decide which pictures I was going to post to Pics4Sharin, the picture sharing site.***

Right at this moment, warning flags should be going off in your head. You lose control over anything that is posted on the Internet. Anyone can download or copy what you upload, and in a case like this, it means they possess your picture. Do you want strangers having your picture on their computer? If you're not sure, ask yourself this question: If a stranger walked up to me on the street and asked me to pose for a picture, would I? Of course you wouldn't. But if you post on open Internet sites it's the same thing except you're sharing with many more than just one person on a street.

Also, once something is posted to the web you can consider it permanent. If several people download a picture or video that they think is funny, they own it. If for some reason the site they downloaded it from disappears, they still have the picture and can re-post or send the picture to whomever they want. THE PICTURE NEVER GOES AWAY!

***I made sure I had pictures that included all of the team. I even made a list of each of my teammates then checked them off once I'd found at least one good picture to put up.***

Posting pictures of friends, teammates, or classmates without their knowledge or permission is not a good thing to do. You are sharing personal information about someone else. Even though it may seem okay because you're not sharing their name or addresses with the world, a picture is a big piece of personal information that should never be shared without someone knowing about it or without receiving their permission. David feels terrible about what happened to him. How bad would he feel if they chose a friend's picture that he posted instead of his? He would feel terrible that he put his friend through this.

***I opened my account and clicked on Hawk's Basketball. It was a group site where anyone could upload pictures.***

Anyone can upload. That also means anyone can download. Students are going to do some things, even when they have heard all the rules about safety. If you are posting anything on a "public" site or page, you should ask yourself three questions.

One: Is there any chance I could regret this later?

Two: Can this picture harm my reputation?

Three: Do I mind if everyone in the world sees it?

If picture sites are going to be part of your life on the Internet, the sites should be private. When they are private you have a little control, although not total control. Your site should be password protected and the only people who have access are allowed by you. Even with this type of safety feature, it is still not 100% safe.

Consider this: you see a picture on a private site and think it's the best picture you've ever seen. You know your friends will like it. The problem is they don't have access to the private account. If you share the password, you may start a chain reaction of sharing. Others might share the password with people you don't know or they'll download the picture and forward it to whomever they want. Just by sharing the password, the account is no longer private. You don't have the right to share someone else's password.

There are so many lessons that can be learned from the comments that are posted. This can easily turn into another cyber bullying issue. Think about how David is feeling. Let's review some of the comments.

> ***WldnCrZyl (12:16 pm): Hey look, it is our school cheerleader captain. Quit the man sports dude, hey I thnk thy hve a sewing club at school, u culd b prezident LOSER***

These kinds of comments don't just hurt David but can also be insulting and hurtful to other people. If there is a sewing club, they have done nothing wrong and have every right to do what they enjoy without being ridiculed. Cheerleading is a sport that requires a lot of talent and abilities, yet they are now victims of torment because of inappropriate comments. We all know this old saying: "If you don't have something nice to say, don't say anything."

> ***Sk8rgrl64322 (1:06 pm): Come on guys knck it off. Take this down***

Although "Sk8rgrl64322" was trying to help, is she? If a victim of this type of behavior sees people responding, does it help? Put yourself in David's place. The more comments there are means the more people know about this and are reading it. Posting more comments, even if good intentioned, may not be the best decision. There are ways to help.

Almost every site that allows picture sharing/social networking allows you to "report" inappropriate behavior. Each site will have a different way to make the notification but almost all have some format to use. This is one way to help. Another way is to get your parents involved. David is feeling alone and confused right now; he NEEDS support and help. Parents can help.

Parents, if your child comes to you and tells you someone is being harassed on the Internet, there are appropriate steps you can take to help. First, talk to your child and praise him/her for getting you involved. Then intervene.

One intervention option is if you know the child who is being bullied, reach out to his/her parents and let them know what is going on. It's a tough phone call to make, maybe, but wouldn't you want to know if it were your child?

Another option for intervening involves the school and police. Your first thought might be "this is not a school issue", and maybe you're right, for now. The next day, this incident will probably be a hot topic in the school, and the school administrators should be advised so they're prepared to deal with it. Some school districts may even consider this a school issue since people who post go to the same school as the victim. The police should also be advised. Will law enforcement take action on every situation? Possibly not, but if it's discovered who posted the pictures, this could be valuable information for law enforcement to have.

> ***What?????? No way we want him on team. We dnt need some guy grabbin us in the shower. Met him at camp, he not that good anywy. Gona show coach these, maybe he could be a cheerleader 4 us instead. Looks like he can dance and hang with the girlz.***

Think about this - you have a dream regarding something that you want to do. You work hard and do everything you can to prepare for your chance to live that dream when someone says, "No way!" Wow! That won't feel good. Trevor was way out of line with his comments when he said that he was going to tell the coach. Trevor should have just kept his thoughts to himself or even better, realized not everything on the Internet is true.

Maria is a true friend. She knew the e-mail she sent David would be difficult for him, but she sent it anyway. Unlike the comments that were posted on the picture sharing site, Maria sent an e-mail, something private that only she and David could read. She didn't post a message where everyone can add comments and make the situation worse. When someone is feeling low, it's good to have a true friend step up and do the right thing. This may give David some strength he needs now to stand up for himself and get help.

As with the incident with Jocelyn and Patti in chapter 4, when things happen on the Internet they never go away. People who get bullied often feel powerless and have no idea how to end the situation.

When others get involved in the proper way, these situations can be resolved. These other people could be parents, relatives, teachers, coaches, clergy people, a family friend or even a trusted neighbor. Many bullying situations take effort from both parents and children to make sure they are resolved in a safe way. This is a crucial time to have good communication. If you or someone you know is being bullied, get help.

# ISSC Answer Key

1. e
2. e
3. a,b,c,d,e
4. a
5. a,b,c,d
6. c
7. b
8. a,b,d

# CHAPTER 6

## Lindsay's Lack of Judgment

**Profile:**
13 year old seventh grader
Mother: Real Estate Agent
Father: Physical Therapist
Siblings: Craig - 17
Aaron - 7
Town demographics: 25,000 residents,semi-rural affluent bedroom community to a medium-sized urban area

My parents say hindsight is 20/20. What they mean is that after something has happened it's easy to look back and see what you did wrong. That's definitely how it is with me.

I'm one of the "3-Gs". My friends and I came up with that name thanks to my dad. My dad always picks on us and calls Alyssa, Molly and me the "girly girls". He says we're girly girls because of the things we love to do. We love being cheerleaders for the Wildcats football team. We love hanging out at the mall and going into the stores to pretend we're buying the jewelry and special purses and cool clothes. We love talking to each other on the phone and IMing on line for hours. Most of my friends love the same things, too, so that's why we call ourselves the "Girly Girls Gang" or the "3-Gs". We're the only ones who know what the three "Gs" stand for. Now I think one of those "Gs" could stand for gullible.

When I turned twelve my parents allowed me to stay home alone for more than an hour. Before that I always felt like someone was "babysitting" me and my younger brother, Aaron. Even when it was just our older brother, Craig, staying with us, I still felt like I was being treated like a baby. Now that I'm older, lots of times I have to babysit Aaron, but I don't get paid for it. That's cool, though, because having the responsibility and knowing my parents trust me makes me feel more grown up.

I remember the day all of my trouble started. Mom, who's a real estate agent, was headed out to a showing. She told me to finish my homework and then I could have the computer until she got home. To most kids, getting computer time might not seem like such a big deal, but my mom works from home, so she's on the computer a lot for work.

I did the math worksheet and finished revising the English paper. So far, seventh grade was easy, and I earned good grades. I'm sure that's why my mom allowed me to have extra on-line time. She said I was more mature. For my twelfth birthday, she even turned off the parental controls as one of my presents. Most of my friends never even had parental controls put on, and it was embarrassing when we'd go on our computer when they came over. It always turned out they'd want to go to some site and I wasn't able to because Mom had it blocked. I'm happy that time's behind me.

I snapped my homework into my binder then glanced at the clock on the microwave. It took me longer to finish than I'd expected. I had half an hour before Mom would be back, but it was time to have some fun.

The first thing I wanted to do was see who was on-line. I clicked on the instant messaging, and right away the screen names for a bunch of my friends and other kids in my class filled the buddy list. My best friend, Alyssa, was on-line. No surprise there. I think she even eats her dinner in front of the computer. Her screen name is *PomPomChick*. She's on my cheerleading squad for the youth football team. Even though we see each other in school and then sat together on the bus an hour and a half ago, we always have plenty to talk about. I messaged her first.

***cheercrzygrl : sup!***

It took a few minutes before she answered. The thing I like about instant messaging is that you can have a conversation with someone *and* listen to music or watch TV or whatever at the same time, so I did a little surfing on the Internet while I waited. After a couple of minutes I heard the computer generated voice say, "*holla*", and I knew Alyssa was there.

***PomPomChick: Hey***

***cheercrzygrl : wht u doin***

***PomPomChick: chatting w/ Greg and some other peeps***

***cheercrzygrl : No Way!!!!***

***PomPomChick: yup***

***cheercrzygrl: where***

***PomPomChick: teencity jarods there 2 ;-)***

My heart jumped. I'd been crushing on Jarod since fifth grade and Alyssa knew it.

***cheercrzygrl: r u chtting w Jarod 2?***

***PomPomChick: yup, he's sportsguy33.***

Jarod's screen name wasn't a surprise since his football jersey number is 33.

***PomPomChick: com here so we can chat n u can chat w him***

I'd heard about the *teencity* chat room, but I'd never been there because of the controls Mom had put on the computer. She told me not to go into chat rooms, but I was curious. My friends visited different chat rooms all the time and talked about the cool people they chatted with. I wanted to see what it was like.

Even though I knew I was alone, I glanced over my shoulder. Maybe this one time I could just go there. I wouldn't stay long, and besides, as Mom and Dad said, I'm more mature now. I'd be careful.

I looked through my IM program and found the link to "chat". When I clicked on it a window opened and there were a ton of chat rooms listed. I couldn't believe some of the names of these rooms and how many people were in them. Kinda freaky. I found the room *teencity*. I was a little nervous, or maybe excited, but I took a breath and entered.

I looked through the list of screen names. There were thirty or more names of kids in the chat. A lot of the screen names were unfamiliar, but I figured they were all kids from my middle school or friends of theirs. Finally, I saw Jarod's. I tried to follow the conversations about school, teachers, hanging out at the mall and subjects like that.

Jarod chatted about football. He'd made a great play in the last quarter of Saturday's game that kept the team I cheer for from scoring. Even though we lost because of it, I wanted to talk with him so badly I didn't care.

As eager as I was to chat with Jarod, I was also nervous because I'd die if he ignored me or didn't care about what I had to say. Since Alyssa was in the same chat room she'd see it and I'd be embarrassed. At first I thought about sending him a private message, but I didn't want to creep him out, so I decided to make a comment and see if he would reply. I waited for the right moment or comment to take a chance.

***footballrulz21: Yeh good tckle but I thnk u got lucky LOL***

***sportsguy33: Whatev, you seem prty tuff on the comp, did not see u out there LOL***

I saw my opportunity to score some points with Jarod. I gripped the mouse and clicked on the text box at the bottom.

***cheercrzygrl: wuz best tckle I evr saw ☺***

A few moments went by and neither Jarod, nor the other kid, responded. I hoped no one else in the room noticed. Heat raced up my face while I waited for what seemed like forever. Then finally, he replied.

***sportguy33: Thnx, ☺ u were at game?***

Chills ran up my spine and my fingers got all jittery. I couldn't believe what was happening. This chat room allowed me to talk to the guy I've been crushing on for the longest time. This was unreal.

***cheercrzygrl: yup u rocked Awsum tckle ☺***

***sportsguy33: thanx who r u?***

***cheercrzygrl: lindsay. i cheer 4 the wildcats***

***sportsguy33: ya sry we beat u ☺***

***cheercrzygrl: lol ya, I know I shuldn't be tellin othr team u were good shhhh don't tell my squad***

***sportsguy33: Lol im good w secrets no prob. How lng u been cheering***

***cheercrzygrl:2 yrs. started when I wuz in 5th grd***

***sportsguy33: ur in 7th***

***cheercrzygrl: explorer team at perry middle***

***sportsguy33: cool. we play u next sat. u should say hey so I knw wht u look like. Wht do u look like? U got pic?***

***cheercrzygrl: no pic, mom would freak. I hve lt brn hair uzually in pony tail***

***sportsguy33: Lol idk wut a french braid is so youll have 2 show me u sound cute***

***cheercrzygrl: thnx, ur prtty cute 2 ☺***

***sportsguy33: I g2g, hope u say hi at game. Cya!***

***Cheercrzygrl: I will***

I was totally psyched that Jarod sounded interested in me. I couldn't wait to tell Alyssa. I was going to close out of the chat room when I heard the tone telling me I'd received an instant message. I checked out the message but didn't recognize the screen name. It made me a little nervous, but I opened the message anyway.

***Speeddemon: hey saw your message u were at game Sat?***

***cheercrzygrl: yup, who this***

***Speeddemon: Steve - you?***

***cheercrzygrl: Lindsay. Do I knw u?***

***Speeddemon: I dont think so, by your screen name I see u cheer, were u cheering at game?***

***cheercrzygrl: yup. Who r u?***

***Speeddemon: I told u, steve which team were u cheering for?***

***cheercrzygrl:wildcats***

***Speeddemon: very cool I must have seen u then, I was there watching***

***cheercrzygrl: u go to my school?***

***Speeddemon: no but my niece goes to the other school, the one u guys were playing***

***cheercrzygrl: niece how old r u?***

***Speeddemon: 25 u?***

***cheercrzygrl: in 7th grade, 13***

***Speeddemon: very cool, u guys were great cheering, blew my nieces school away with your cheers, u guys were much louder and better ☺ sounded better than a lot of high school cheerleaders I have seen***

***cheercrzygrl: wow thnx we work hard. I luv cheerin***

***Speeddemon: hard work show u guys ROCK!!!! ☺***

***cheercrzygrl: ☺***

***Speeddemon: looked up your profile, says u like to shop n cheer, which u like best LOL***

***cheercrzygrl: LOL both***

I couldn't believe a 25 year old guy wanted to talk to me. Me? Usually the guys talk more to my friends than me, but not this time. Steve was talking only to me, not even in the room. This was private between just us. In some ways, it weirded me out, but it was exciting, too.

***Speeddemon: parents there? U alone?***

***cheercrzygrl: mom will brb she just ran out***

***Speeddemon: she know u chat on here***

***cheercrzygrl: NO WAY!! shhhhhhhhhhhhhhhhhhhhhhhhhhhh***

***Speeddemon: LOL never tell, u seem very cool, don't want u in trouble***

***Speeddemon: so what else u like to do beside shop and cheer, and talk to older guys here LOL***

***cheercrzygrl: I duno lots I guess, hangin w friends, mall, movies, nrmal stuff I guess. U?***

***Speeddemon: same except mall LOL love movies, music and hanging with cool people***

***Cheercrzygrl: Wht music u like?***

***Speeddemon: rock, but will listen to almost anything. U?***

***Cheercrzygrl: rock too, dnt like country LOL***

***Speeddemon: you got a picture?***

***Cheercrzygrl: no u?***

***Speeddemon: yes, but I only trade. When u get 1 u can have mine***

***Cheercrzygrl: k***

***Speeddemon: u seem cool, very mature. We should hang out some time.***

***Cheercrzygrl: realy?***

***Speeddemon: Yeah! Would be fun. I have to go, Love to chat with u again soon. You rock Go wildcats LOL***

***cheercrzygrl: k***

**(to be continued in next chapter)**

# I.S.S.C.
# Internet Safety Savviness Challenge

Most times we don't think about what personal information we're giving away in ordinary conversations on-line. However, an "Internet predator" is looking for these slips so he or she can find out as much about you as possible.

In the chat room, this person, *Speeddemon,* was lurking, waiting and gathering enough information to contact Lindsay and make it seem like he knew her. This made Lindsay feel more comfortable chatting with the stranger and willing to take a risk she might not ordinarily take.

Do you think you're more Internet savvy than Lindsay? Review the conversations Lindsay had in the chat room with Jarod (*sportsguy33*) and also with *Speeddemon.* Can you identify the personal information Lindsay revealed during both chats? Make a list of the personal information Lindsay accidentally shared during her on-line conversations. Also, identify the mistakes her parents made.

(Answer key available at the end of the chapter.)

# What Can We Learn?

> ***She told me to finish my homework, and then I could have the computer to myself until she got home.***

The first thing that went wrong was Lindsay's mom got soft on her rules. If parents allow kids to use the Internet while being supervised, it's one thing, but when kids are allowed to use the Internet while there is no parent home, it's not a good strategy for safety. Things happen on the World Wide Web at a very fast pace. Due to this fact, things can happen or get out of control quickly. Children should have a parent present so if something does go wrong it can be looked into quickly to ensure the safety of the child.

> ***When I turned twelve she even turned off the parental controls.***

Parents sometimes want to be their children's "buddies". With the Internet, this is one time we must remain parents. Kids, although they may show signs of "maturity", are still kids with a lot to learn. The Internet is not a place for this education. Parental controls are a big key in keeping children safe. It would be easy for parents to fall for the "none of my friends have parental controls" or the popular "don't you trust me?", but don't do it.

First, it doesn't matter if other children don't have parental controls, it just means that they aren't as safe as they should be. Yes, we do trust our children - if they have earned it. But, even though we trust them, parents must realize how much danger is out there for children.

Studies show that the average age of first exposure to child pornography is 11 years old. 11 YEARS OLD! This doesn't usually happen because children are looking for pornography; it can happen just by clicking on the wrong site or misspelling a web site address as we saw in *Andrew's Accidental Search*. Parental controls can not only assist in preventing our children from communicating with strangers, they help prevent innocent mistakes. Unfortunately, with technology, when kids make these types of mistakes, they are exposed to things that no children should see.

***She told me not to go into chat rooms, but I was really curious.***

Rules for using the Internet are very important and children need to understand this. Lindsay's mom had set up rules probably because she knew the dangers that went along with chat rooms. Lindsay knew her mother would not approve, and it was against her mother's rules, yet she still went into that chat room. If this rule was never violated, Lindsay would not be getting herself into a scary situation.

***I couldn't believe some of the names of these rooms and how many people were in them. Kinda freaky.***

If you go into an area, whether it's on the Internet or in the real world, and you think to yourself, "kinda freaky", you need to get out of that area. When Lindsay opened up the chat room listing and saw the available chat rooms and kind of got freaked out, she should have stopped right there and left this area of the Internet. Her instincts told her that she was making a poor decision by going there. Chat rooms allow strangers to communicate with one another, and we all know that can lead to trouble. Also, a chat room titled "*teencity*" will attract teenagers to it, but it will also attract people who *like* teenagers.

Parents have to understand that chat rooms have very little security and few, if any, rules. Some have "administrators", but these people can only kick someone out of a certain room. People can always "report" bad behavior to the site hosting the chats, but this isn't foolproof, either. If a person gets kicked out of a chat room because of inappropriate language, they log off and come right back using a new screen name. Security is very limited. Another thing that parents may not understand is that some sites that host chat rooms allow members/users to create topics. These chat room topics have no boundaries and can be about any topic under the sun, and yes, that includes sex and drugs. Children can gain access to these rooms just as easily as any adult.

> ***I checked out the message but didn't recognize this screen name. It made me a little nervous, but I opened the message anyway.***

No matter where you are, a chat room, game room, social networking site or instant messaging, if you receive a message from someone you do not know you should not accept the message. Children are taught at great lengths when they are younger about "stranger danger". This message is so driven home that you can ask any kids about "stranger danger" and they will say, "I know I should never talk to a stranger or take anything from a stranger." But on the Internet this common sense approach to safety is often ignored. We all feel safe in our homes, as we should, but when we connect to the Internet and talk to strangers we are not just in our homes, we are out in the world. All stranger safety rules that apply in our neighborhoods should be used on the Internet. If you would not talk to a stranger in your neighborhood, what makes you think it is any safer to do this on the Internet? A stranger is a stranger no matter where you make contact with him or her.

> ***Speeddemon was talking only to me, not even in the room. This was private between just us. In some ways it weirded me out, but it was exciting, too.***

Children need to learn to follow their instincts. "It weirded me out" should be followed by "I stopped communicating with him". No one who is 25 years old should be having conversations with kids who are in their early teens, especially private chats. There are many adults on the Internet, millions, and a 25 year old man can find one of them to talk to, NOT A 13 YEAR OLD. Children need to not be afraid to say no or just end communications. We know that kids are taught to respect their elders, but if it's a situation that's unsafe or makes a child feel uncomfortable, he or she has every right to stop this communication.

***Speeddemon: looked up your profile, says u like to shop n cheer, which u like best LOL***

No teen or child should have a profile. A profile is a quick reference for strangers to find personal information on you. Whatever you put on your profile, anyone can see. If you provide likes, dislikes, hobbies, birthday, name, school you attend and any other information, you are an easy target for someone with bad intentions who wants to "friend" you. You do not need a profile.

Lindsay can teach us all a lesson. She is a good kid and, for the most part, follows her parents' rules. She knew what she was doing was wrong, but she was lured by the excitement of something different and new. If she had followed safety rules that she had learned in kindergarten, DON'T TALK TO STRANGERS, none if this would have happened.

Chat rooms are rooms full of strangers. One second of poor judgment opens up a world of danger. Now that she started talking to this stranger, chances are very good that this could lead to even bigger problems.

# ISSC Answer Key

**Lindsay's mistakes:**

- gave her age
- gave her grade
- gave her first name
- revealed she's a cheerleader
- told what sports team she cheers for
- gave a physical description of herself (light brown hair, French braid)
- admitted she was home alone
- revealed the name of the school she goes to
- told what "team" she's on at school
- offered when and where she'll be cheering the next Saturday
- She didn't follow her parents' rules. She's not supposed to be in chat room
- She won't tell her mom she was experimenting on the Internet

**Parents' mistakes:**

- Lindsay was allowed unsupervised time on the computer
- Lindsay's parents turned the parental controls off
- allowed Lindsay to use a screen name that revealed personal information

# CHAPTER 7

# Lindsay's Lack of Judgment – Continued

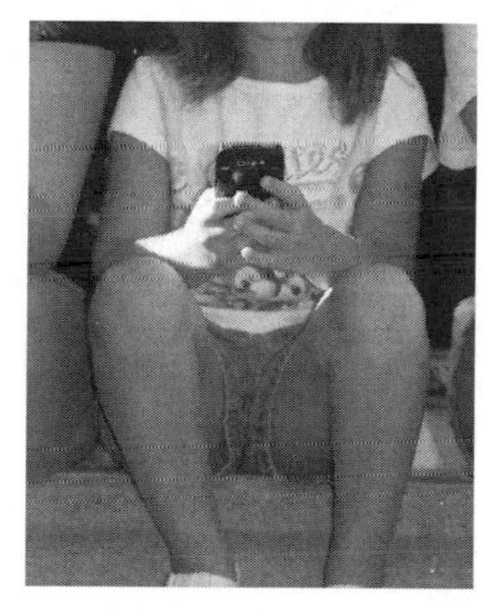

I rolled over in my bed, stretched and checked the time. 8:30 ! It had been a crazy couple of weeks with cheerleading at football games, homework, and a dance at school last night.

The bathroom door closed downstairs and then the shower started. I knew Dad was golfing, Craig was at work, and I doubted Aaron would be getting into the shower on a Saturday morning, at least not without pitching a fit as he battled Mom, which I'd definitely hear, so that meant it was Mom. Which also meant the computer was free, and no one would be hovering over my shoulder.

I raced down the stairs. Aaron was planted in front of the television watching cartoons. He was such a zombie when those were on; he'd never even know I was up. I couldn't wait to talk to some of the girls about last night's dance. It was our first one, and it was awesome.

The room where we had the family computer doubled as the homework room and Mom's office. The computer was already on, and a real estate screen was up. I minimized it and clicked into my instant messaging program. I scanned down my friend list hoping Alyssa was on, but she wasn't; however, Molly was. A message from her popped onto my screen immediately.

***purrfexion021499: hey linds ur up early***

***cheercrzygrl: ya***

***purrfexion021499: dance last nite wuz kewl huh?***

***cheercrzygrl: ya***

***purrfexion021499: Alyssa wants 2g2 c a movie l8r***

***cheercrzygrl: ya I know***

***purrfexion021499: you going?***

Before I could answer, another IM came up. It was *Speeddemon*. Apparently everyone was up early today.

***Speeddemon: HEY!!!***

I clicked back to Molly's message box.

***cheercrzygrl: brb***

***purrfexion021499: k***

I hated to ditch Molly, but I could talk to her any time. *Speeddemon* isn't on all the time like my friends. I tried to add him to my friend list a couple of days ago but for some reason it didn't work, so I have to wait for him to IM me. I clicked on the little Tasmanian devil next to his screen name.

***cheercrzygrl: hey***

***Speeddemon: Wht up*** ☺

***cheercrzygrl: nutin wht up w u?***

***Speeddemon: been waiting for you. How r u? still cheering?***

***cheercrzygrl: of corse alwys cheerin?***

***Speeddemon: are you alone?***

***cheercrzygrl: kinda***

***Speeddemon: ?***

***cheercrzygrl: mom in shwr***

***Speeddemon: oh too bad ur not alone***

***cheercrzygrl: y?***

***Speeddemon: wanted to call and hear ur voice, saw you cheer, want to hear you***

***cheercrzygrl: WOW dunno***

***Speeddemon: what?***

***cheercrzygrl: kinda crazy***

***Speeddemon: that is your middle name cheerCRZYgrl, LOL it would be cool***

***cheercrzygrl: lol I gues***

***Speeddemon: you don't wnt to talk to me?*** ☹

***cheercrzygrl: ddnt say tht jst seems I dunno***

***Speeddemon: exciting*** ☺

***cheercrzygrl: kinda scary 2***

Someone walked into the room and I almost jumped out of my skin. I hit the minimize button on the screen and swiveled in the chair praying Mom wasn't standing behind me. No. It was Aaron in the doorway, his pajamas all twisted around his waist and his short brown hair sticking up in every direction.

"What do you want?" I snapped.

"Geez, what a grump! I just wanted to know if you're gonna use the rest of the milk?"

My blood pounded so hard in my ears that his words were almost muffled. "What do you care?"

"'cause I want to have some more cereal, but Mom said I had to ask you first."

I swear every nerve in my body had to be jumping. I just wanted to get rid of him. "Go ahead and have the milk. I'll have waffles."

"Yes!" He fist-pumped the air as if he'd won some great victory then dashed toward the kitchen.

When I turned back to the computer I was greeted by the screen wallpaper which was a big family picture. My mom stood right behind me in the picture. A big smile lit her face and her eyes seemed to be watching me. I stared at it for a second then remembered *Speeddemon* was waiting. I pushed the space bar to bring the regular screen back.

***Speeddemon: u still here***

***cheercrzygrl: ya***

I wasn't sure what to say now. I never had some guy say he wanted to hear my voice before. It seemed kind of weird, but then I wondered if maybe older guys said stuff like that all the time. I didn't want him to think I was a baby.

***cheercrzygrl: hve 2 think bout talk***

***Speeddemon: ok I get that but I am not a freak!! just think u r super cool***

***cheercrzygrl: really, thnx u seem vry cool toooooo***

***Speeddemon: I am lol***

***cheercrzygrl: ☺ ur funny***

***Speeddemon: don't forget cool LOL***

***cheercrzygrl: LOL cool2***

***Speeddemon: ☺ you rock***

This flirting thing was new to me, but I had to admit I liked how it felt. I'd seen enough of it in the movies to pick up on how to do it. It made me feel more mature, especially since I was flirting with some dude who was 25. I still couldn't believe he was interested in me, but he seemed so cool. What could it hurt to flirt with the guy over the Internet? I wanted to sound casual like this was something I do all the time, so I took a deep breath and decided to control the conversation.

***cheercrzygrl: wht u up 2 2day?***

***Speeddemon: not sure maybe just hang w some friends u?***

***Cheercrzygrl: movies l8r I hope nt sure***

***Speeddemon: movies cool, where do u go***

***Cheercrzygrl: mall***

***Speeddemon: you go to a mall, NO WAY!!!!!! LOL***

***Cheercrzygrl: u mking fun of me? Lol***

***Speeddemon: LOL never, I like u too much to make fun of u***

***Cheercrzygrl: OHHHHHHHHHHHHHHH ur sweet***

***Speeddemon: you are sweet***

***Cheercrzygrl: wher u go 2 movis Mall?***

***Speeddemon: I cuuld, are you inviting me? ☺***

***Cheercrzygrl: I ddnt say tht wuz jst askin***

***Speeddemon: ohhhh sniff sniff ☹***

***Cheercrzygrl: no wait I ddnt mean to hrt ur feelin sorry***

***Speeddemon: LOL relax jk***

***Cheercrzygrl: phew!!!!! tht freaked me out lol***

***Speeddemon: think it is sweet that you got upset thinking you upset me, means you like me***

***Cheercrzygrl: tht wht u think***

***Speeddemon: yup***

I stopped typing and sat back in the chair. Even though I knew I should sign off, or at least go back to Molly, I had to admit I *was* starting to like this guy. Not only did he make me feel more grown up, sometimes it seemed like he knew me better than most of my friends.

Between my conversation on-line and the high-pitched voices of the animated characters in Aaron's cartoons, I was distracted and hadn't noticed when the shower was turned off. Even though I didn't want to stop chatting, I'd have to be off here before Mom came out. She'd flip if she knew I was chatting with a guy, especially an older one. I sat forward in the chair and put my hands back on the keyboard.

***Cheercrzygrl: hey i g2g***

***Speeddemon: so u didn't answer me bout the movies does that mean I am invited?***

***Cheercrzygrl: I dunno u realy wnt 2 go w me?***

***Speeddemon: OH YEAH!!!! Been thinking bout u since I saw u at Sat's game***

***Cheercrzygrl: u wer there? knw u wer there 2 games ago ddnt knw u came to last 1***

***Speeddemon: yup after we chatted here, you seemed cool, had to see you again. Have not stopped thinking bout you since***

***cheercrzygrl: WOW***

***Speeddemon: what? You all right?***

***cheercrzygrl: yeh jst no 1 ever said nething r don nething like that 4 me***

***Speeddemon: what? tell u how cool you are and go watch u at ur game***

***cheercrzygrl: nt tht I dunno***

***Speeddemon: what?***

***Cheercrzygrl: u mkin me smile ☺ u r vry cool***

***Speeddemon: I think you are very special ☺***

The knob on the bathroom door squeaked so I knew Mom was coming out.

"Aaron, the t.v. is kind of loud," she said. "Could you please turn it down before you wake up your sister?"

"She's already up." The volume on the television suddenly increased. "She's glued to the computer." My little brother can be so obnoxious sometimes. I honestly don't know why Mom lets him get away with some of the things he says and does.

"As soon as I'm dressed we have errands to run." Her voice faded in and out which told me she was probably poking her head into the kitchen and living room. I was afraid she would come looking for me, so I knew I had to get off the computer fast.

***cheercrzygrl: I g2g help mom 4 bit, u gonna be here l8r?***

***Speeddemon: if you are going to b here I will b here ☺ bye sweetie xoxo***

***cheercrzygrl: bbl cya☺***

I clicked out of the screen and then put Mom's real estate page back up. I was feeling jittery as I thought about *Speeddemon.* He called me sweetie. Other than my parents and grandparents, no one had ever called me that. I'd heard guys say things like that in movies and on television, but it was different having some guy say it to *me.* I pictured myself out somewhere with him with his arm around me. It seemed romantic. *Speeddemon* seemed romantic. I couldn't wait until we got back home so I could go on the computer again.

The errands and lunch took forever, and I thought we'd never finish. Several times I glanced at the clock wondering if *Speeddemon* was waiting for me to sign on.

After I swallowed a bite of tuna sandwich, I said to Mom, "I have a social studies project to work on. Can I have the computer for a while?"

"What's the project on?" She picked up Aaron's and her plate and carried them to the sink. I was glad because with her back to me she wouldn't be able to see the guilty look I figured was on my face.

"We're studying South Africa. I want to get some information on Nelson Mandela." It was true, we were studying South Africa, but there really was no reason for me to look up information on Mandela. We'd been reading books and articles about him for a couple of weeks, and I felt like there wasn't much more I could learn. A website about him would be a good cover page, though.

Mom turned on the water in the sink and began rinsing off the plates. "I have some work to do, too. We'll have to share the time."

The phone rang and I imagined a 25 year old guy on the other end. I sprang to answer it as if I thought he really would call. When I heard Mom's friend, Lucy, on the other end I was actually disappointed. I handed Mom the phone as a thought buzzed through my head. What would I have said and done if it *had* been *Speeddemon*? The idea terrified me, but at the same time, I was curious to know what he sounded like, too. This whole situation with him was getting me pretty confused, but it was a fun confusion. I figured all teenagers probably felt this way sometimes when they met someone new.

With the phone to her ear, Mom picked up her iced tea and moved to the living room. That told me she was settling in for a long chat. Inside I did a little cheer. She'd be busy for a while. I couldn't believe my luck.

I put my sandwich plate in the dishwasher, grabbed my yogurt and spoon from the table and went into the den. I dropped into the chair, set the yogurt on the desk and signed on. My backpack was by the door in the kitchen, so I hurried to get it. I had to at least make it *look* like I was doing homework.

Back at the desk I opened the social studies book to the chapter on South Africa and set it next to the computer. A twinge of guilt about being sneaky flickered in my stomach, but I knew Mom would be less likely to come in and check on me if she thought I was doing homework.

I settled into the leather office chair and scooted closer to the computer. First, I put in a search for information about Mandela and quickly came up with a page that would work if I needed a cover. I minimized that and logged onto Instant Messaging.

***Speeddemon: bout time***

***cheercrzygrl: lol wht u mean***

***Speeddemon: been waiting for you. how was lunch? is your belly full LOL***

***Cheercrzygrl: yup, tuna YUM, helpd mom, now "doin homework"***

***Speeddemon: homework yuck!!!!***

***Cheercrzygrl: shhhhhh nt rely, mom thnks I am doin that now, shhhhhhhhhhhh***

***Speeddemon: sneaky lol I like that see you** are **crazy!!***

***Cheercrzygrl: LOL mybe***

***Speeddemon: so are you going 2 movies***

*Cheercrzygrl: thnk so prb jst me n Alyssa but mbe Molly 2*

*Speeddemon: who that*

*Cheercrzygrl: my bff been friends 4ever*

*Speeddemon: cool, they cheer too*

*Cheercrzygrl: yup*

*Speeddemon: cool, what movie are you gonna see*

*Cheercrzygrl: n idea, we pik when we gt ther. Y?*

*Speeddemon: can I ask you a question*

*Cheercrzygrl: yup*

*Speeddemon: you got a bf?*

*Cheercrzygrl: no u gt a gf*

*Speeddemon: nope why dont you have one*

*Cheercrzygrl: jst dnt u?*

*Speeddemon: I did have gf but dont anymore*

*Cheercrzygrl: oh sory*

*Speeddemon: oh no problem, she was weird anyway lol*

*Cheercrzygrl: o wht mde her weird*

*Speeddemon: long story, not as nice as i thought she could have*

*been. stopped having fun, she got old quick LOL*

*cheercrzygrl: gt old, she sik*

*Speeddemon: LOL no not sick she just did not know how to have fun, she got old like that see*

***cheercrzygrl: oh duh, gt it so she nt a granny lol***

***Speeddemon: LOL nope my age, that is why I like you, you seem fun***

***cheercrzygrl: lol im nt old***

***Speeddemon: I know ☺ can I ask you another question***

***cheercrzygrl: yup***

Mom's voice was louder. She was coming toward the den.

***Cheercrzygrl: wait brb***

I moved the cursor up to minimize the chat then quickly clicked on the icon to bring up the site about Mandela. Mom's footsteps grew louder, and then she was in the room coming right toward the computer. I stared at the screen, pretending to read. Engrossed in conversation, Mom rummaged through a file on the desk until she pulled out a couple of papers. She was telling Lucy something about the acreage on a listing she was handling.

A jolt of fear shot through me. I couldn't remember if I'd turned off the sound on the computer. What if *Speeddemon* sent me a message and Mom heard the low growl that went with his screen name? I moved the mouse on the pad and clicked on another section of the website. I didn't dare look up at Mom. My face felt so hot I was afraid it was bright red and would give me away. I read the same sentence on the website ten times before Mom finally picked up the whole file and left the room. My lungs burned when I exhaled. I hadn't realized I was holding my breath.

I glanced over my shoulder to be sure she was gone then turned back to bring up the instant messaging screen. I swore my face and hands were on fire.

***cheercrzygrl: srry im back***

***Speeddemon: everything ok?***

***cheercrzygrl: ya jst had 2 do smthing wht wuz ur question***

***Speeddemon: oh, yeah, almost forgot LOL you ever make out with a guy***

***cheercrzygrl: guy? U mean oldr***

***Speeddemon: any***

***cheercrzygrl: yup***

***Speeddemon: was he your age***

***cheercrzygrl: yup***

***Speeddemon: you like it?***

***Cheercrzygrl: yeh I gues***

***Speeddemon: only guess, he must not have been a good kisser LOL***

***Cheercrzygrl: LOL nt rely***

***Speeddemon: did you guys make out in the movies, sit in the back so no one can see?***

***Cheercrzygrl: LOL no***

***Speeddemon: where***

***Cheercrzygrl: skool dance***

***Speeddemon: oh I see, u never make out in movies***

***Cheercrzygrl: nop, u evr make out there***

***Speeddemon: movies, yeah I like it, dark people around but cant see, very cool***

***Cheercrzygrl: gues, sems cool***

***Speeddemon: it is, you should try it***

***Cheercrzygrl: LOL yeh me n ALyssa LOL dnt thnk so***

***Speeddemon: not with her***

***Cheercrzygrl: thn who***

***Speeddemon: me ☺***

***Cheercrzygrl: NO WAY, r u serius***

***Speeddemon: yeah why not, you said I seem cool, I know you are cool***

***Cheercrzygrl: WOW***

***Speeddemon: so you wanna meet at movies later?***

***Cheercrzygrl: I duno sems outa control***

***Speeddemon: it could be, that is the fun ☺***

Out of all of my friends, I'm the least crazy one. Sometimes they tease me about being such a rule follower and how I have to "let go" once in a while or I won't have any fun. I'm not really sure what breaking rules has to do with having fun, but I'll admit sometimes I worried they wouldn't want to hang out with me any more because I won't do some things. My friends aren't bad; they're just risk takers and I'm usually not. This thing with *Speeddemon* was definitely not something I normally would do, but it made me feel more like I fit in. Even if the other girls didn't know I was doing it, sometime I might tell them and I'm sure they'd think it was pretty cool. There were still some things I wasn't quite sure about, though.

***Cheercrzygrl: u evr do this b4?***

***Speeddemon: what meet a grl at the movies, sure just no one as cool as you ☺***

***Cheercrzygrl: wre thy my age?***

***Speeddemon: some, I don't care about age I just like havin fun don't you?***

***Cheercrzygrl: yeh, I dunno***

***Speeddemon: oh come on, I can't stop thinkin about you. Your friend will be there so it is not like I am going to take you away, we will just meet at movie and see how it goes***

***cheercrzygrl: I dunno***

***Speeddemon: you know you want to, it is crazy and fun and you want to I know you do***

***cheercrzygrl: how u knw***

***Speeddemon: cause you are my crzygrl*** ☺

For a second I felt funny. This seemed to be happening fast, but then I'd never had an older guy interested in me, so I figured I was just inexperienced.

***cheercrzygrl: mom neds me, g2g, bbl n let u knw cya***

***Speeddemon: think about it it would be AWESOME to finaly meet bye xoxo***

I signed off and stared at the screen. Mom didn't need me. I needed time to think. In one way, I wanted to call Alyssa and Molly and ask them what they thought about all of this. I wondered if they would tell the guy they'd meet him at the movies.

I didn't have a long time to think about it. A few minutes later, Mom asked me if I wanted to babysit for Lucy's three little girls for a few hours. I jumped at the chance for two reasons. One, it would give me time to decide about *Speeddemon* and the movies. Two, I'd be earning money so if I did decide to go, Mom and Dad were sure to let me since I could pay my own way.

The four hours dragged by even though the babysitting went well and I earned $30. That was definitely enough for the movies and maybe a little shopping at the mall if there was time. When I got home Dad was reading the newspaper in the living room and Mom was making dinner. I decided to ask Mom first because Dad would tell me it was up to her.

"Hi, Mom." I crossed the kitchen and went to the cupboard to take out plates and glasses to set the table.

"How was babysitting?" She was stirring gravy to go on the pot roast she'd made.

"Good. The girls are always fun." I knew it was best if I asked her while setting the table because she'd start out happy with me for being responsible. After putting the five plates and glasses down I went to the silverware drawer.

"Alyssa and Molly are going to the movies later and wondered if I could go, too. I'll use my own money."

Mom asked me the usual questions about what movie, how we were getting there, and what time the movie started and ended. I knew if she was asking me all of those questions that she was probably going to say yes, and she did. After I finished setting the table I texted Alyssa and Molly then went in to the computer and signed on to instant message again. Immediately the low growl sound came across and I knew *Speeddemon* was there.

***Speeddemon: been waiting 4 you ☺***

***cheercrzygrl: u stil here?***

***Speeddemon: yup***

***Cheercrzygrl: ur crzy ☺***

***Speeddemon: about you, yup!!***

***Cheercrzygrl: OOOOOOOOHHHHHHHH ur d best***

***Speeddemon: nope you are ☺ xoxo***

***Cheercrzygrl: u ben missin me?***

***Speeddemon: like CRAZY!!!!!!***

***Cheercrzygrl: LOL***

***Speeddemon: you been thinking?***

***Cheercrzygrl: yup***

***Speeddemon: what about***

***Cheercrzygrl: movie***

***Speeddemon: and.......***

***Cheercrzygrl: lol nuthn else***

***Speeddemon: Oh come on you stink! LOL what are you really thinking about***

***Cheercrzygrl: if u r serius?***

***Speeddemon: Very Serious! Think you could handle meeting a guy like me***

***Cheercrzygrl: yup***

***Speeddemon: oh really, you sound pretty confident***

***Cheercrzygrl: nt a baby***

***Speeddemon: never said you were. So is that a yes, im invited?***

***Cheercrzygrl: I dunno, I dnt wnt 2 get in truble***

***Speeddemon: me either***

***Cheercrzygrl: I dnt wnt 2 ditch frinds***

***Speeddemon: they sound cool, theyd understand if you left for a***

***few, or just say you are going to bathroom, we could just meet for a few mins.***

***Cheercrzygrl: sunds cool but nt sure***

***Speeddemon: what are you not sure about***

***Cheercrzygrl: leavin Alyssa and Molly, getin n truble, lotz***

***Speeddemon: if you were with Alyssa and Molly and one of them said, I am going to go say hi to a boy i know, would that bother you***

***cheercrzygrl: no***

***Speeddemon: so I bet they'd be cool with it ☺***

***cheercrzygrl: I gues***

***Speeddemon: and we will not leave theater so don't worry about trouble, so dark no one will know we meet I don't want trouble either.***

***Cheercrzygrl: ur rite***

***Speeddemon: so that is a yes!!***

***Cheercrzygrl: mmmmmm how is this gona work, tell me***

***Speeddemon: you give me your cell number, I text you when I get there, you tell me what movie and I will sit in back, when you say you are going to bathroom, come see me***

***cheercrzygrl: k hw will I know it is you?***

***Speeddemon: good question, you are smart… let me think about that***

***Cheercrzygrl: wht we gona do***

***Speeddemon: say hi, talk for a bit, would love to kiss you in movie, be your first movie kiss***

***Cheercrzygrl: I duno***

***Speeddemon: ok if you don't want to kiss, that cool, but at least we can meet. I can hear your voice and see you up close***

***cheercrzygrl: tht is it***

***Speeddemon: well if you want more, I am up for anything but thought you might want to just meet and maybe kiss. What do you want to do?***

What did I want? I tried to picture me with him. It was one thing to hang out with Justin at the dance last night. He's my age and I didn't feel like he really knew any more about making out than I did, although I wasn't sure a couple of kisses and holding hands was really making out. But this guy had to be more experienced. What if I ended up embarrassing myself because I didn't know what to say or do?

***Speeddemon: did I lose u?***

***Cheercrzygrl: no no, nt sure wht I wnt, we wil nt leave movie rite***

***Speeddemon: right***

***Cheercrzygrl: n if I dnt wnt 2 kiss***

***Speeddemon: no problem, but I think you do***

***Cheercrzygrl: ths is very cool, bt I am nervus, dnt wnt truble or get hurt***

***Speeddemon: no trouble and I would never hurt you, you are my crazygirl***

***Cheercrzygrl: ooooooooohhhhhhhhhh u r da best***

***Speeddemon: you rock***

***Cheercrzygrl: u sur dis wil work***

***Speeddemon: absolutely, nothing to worry about you can come say hi and go back if that is all you want, but I think you want more***

***cheercrzygrl: I duno, stil nt sure I wil knw its u***

***Speeddemon: mmmmmmm got it, I will wear a hat, when I see you walk up, I will take my hat off***

***cheercrzygrl: wil I c that in movie***

***Speeddemon: sure, remember I have seen you so I know who to look for***

***cheercrzygrl: oh yeh, good point lol***

***Speeddemon: it will be exciting, crazy and cool altogether, you know you want to meet me I am cool remember***

***cheercrzygrl: u r cool jst dnt knw u so wnt to b safe that all***

***Speeddemon: you are smart. I wont hurt you, PROMISE***

***cheercrzygrl: Ill scream n movie if I hve 2***

***Speeddemon: I know but you won't have to, you will like me so much youll kiss me***

***cheercrzygrl: I duno***

***Speeddemon: you will, I know it ☺***

***cheercrzygrl: we r gona go 2 7:30 mov***

***Speeddemon: no problem, mall theaters right***

***cheercrzygrl: yeh***

I felt better about this. There were lots of people around in the mall, so I'd be safe. Maybe Alyssa and Molly were right. Maybe it *was* time for me to be a little more daring.

***Speeddemon: very cool, give me your cell and I will text you when I get there***

***cheercrzygrl: 555-1357, dnt giv tht 2 neone***

***Speeddemon: told you I am good with secrets, will never give to anyone.***

***Cheercrzygrl: k, u beter not***

***Speeddemon: I wont trust me***

***Cheercrzygrl: k cya l8r***

***Speeddemon: mcant wait XOXOXOXOXOXOXOXOXO***

***Cheercrzygrl: k***

I pulled my cell phone from my pocket and pushed the shortcut button for my contact list. I loved that my best friend's name started with A. It saved time when I wanted to send a message or call. I clicked on "new text message" and typed on the keypad.

***Im in 4 movie c u at 6:30***

# I.S.S.C.
# Internet Safety Savviness Challenge

Are the statements below true or false? Test yourself and your Internet savviness by taking this quiz. (Answers at the end of the chapter.)

TRUE OR FALSE

1. ____ Meeting in a public place is the safest way to meet someone for the first time.

2. ____ There is no way a person could find me with only my cell phone number.

3. ____ The computer in a common area is a good idea for both kids and parents.

4. ____ When I feel "freaked out" while on the Internet I should just ignore those feelings because it is just on the Internet.

5. ____ No one can ever find out true information about me from a screen name that I create.

6. ____ Screen names should never include any personal information.

7. ____ If you are going to meet a stranger, it is best if you bring your friends with you to be safe.

8. ____When someone on the Internet tells you their name, they have to tell you the truth.

9. ____ There are people who use the Internet to build relationships and meet with young people.

10. ___ It is okay for a parent to check a child's text messages and check what they are doing while on the Internet.

11. ___ If you're on the Internet with someone you don't know, the best thing to do is act older so they can't tell your age.

# What Can We Learn?

***The room where we had the family computer doubled as the homework room and Mom's office.***

Once again, the computer needs to be in a public area of the home. An office may not be the best choice. Kids will say, "I need my privacy," and this is true, at times. If you want to be private, stay off the Internet. Once you connect and start surfing the web, privacy is not an option. Things, both good and bad, happen fast on the Internet and when children are in a "private" area of the house, parents may not be aware of potentially dangerous situations occurring.

***Speeddemon: wanted to call and hear ur voice, saw you cheer, want to hear you***

At this point, *Speeddemon* knows that Lindsay is a girl in 7th grade and is 13 years old. He says he wants to hear her voice? He is 25 years old, or so he says. His request to talk on the phone should send up warning flags that there may be reason to be concerned. Chatting with strangers in chat rooms and instant messaging is bad enough; now add the personal connection of hearing a voice and this becomes very dangerous. Not only is it completely inappropriate for a man this old to want to talk to a 13 year old girl, but you need to think about safety.

With technology like caller identification, if Lindsay said yes and called him, he would then have her home phone number or cell phone. It doesn't take a lot of research to find someone these days if you have their phone number. Phone numbers are one of those items of personal information that you should never give out to strangers, and no matter how cool *Speeddemon* sounds, or how long she's been chatting with him on-line, he is still a stranger.

***Speeddemon: that is your middle name cheerCRZYgrl***

This is a good time to talk about screen names. Think about what your screen name says about you. Lindsay felt it was cool to have "crzy" as part of her name. Don't think for a second that people like *Speeddemon* won't pick up on things like that. They do. Screen names should not reveal any personal information about you such as your real name, address, zip code where you live, phone number, date of birth or the number you wear on a sports uniform. They should also not be things that may attract the wrong type of people. Screen names that include words like "flirty", "sassy", "buff", "sexy", "hot", "wild", "cute" and "crazy" are just asking for trouble.

Let's look at Lindsay's "cheercrzygrl". It looks fun, doesn't it? At first glance you might not think this is that bad, but it could be. If a person is trying to find someone's true identity on-line it is like putting a puzzle together and each bit of information is a piece of that puzzle. When someone puts "cheer" in her screen name, it's likely she is a cheerleader or very cheery. Most common would be the first one. "crzy" says this person is probably a rebel, out of control and is willing to do anything. "grl" implies the person claims to be a female.

Those three alone may not be that bad, but then in chats Lindsay said she was a cheerleader, what school she attended, her age and, she also revealed her real name. By using a screen name like this and answering a few simple questions, that puzzle is already being put together. Screen names should be very generic. For example, if someone used "ABCD1234", not one piece of the puzzle of a person's identity is given away.

Here is a good plan: parents and their children should sit down and create screen names together before any name is ever used. Kids, if your first thought is, 'No way do I want my parents involved in picking my screen name' then your parent should ask, "Why not?" If you are willing to have people on the Internet know you by this name then your parents should know you by this name, as well. If you think the name you want would not meet your parents' approval, you are right and should understand why.

***Besides, what could it hurt to flirt with the guy over the Internet?***

It could hurt a lot. Although at this point Lindsay may not be taking this whole situation seriously, "*Speeddemon*" is. If this type of behavior continues, "*Speeddemon*" may think that Lindsay is very serious, and he would make every effort to find her and make a real life connection with her. Remember, he already knows what school she goes to and what she looks like. It would not take a lot for him to find her again. And probably the most important thing of all, no 13 year old girl should be flirting with a 25 year old male.

***Speeddemon:*** ***movies cool, where do u go***

***Cheercrzygrl:*** ***mall***

This makes no sense. Telling a stranger, "Hey, from 7:15-9:30 here is where I will be" is not thinking about safety. In fact, you are not thinking at all. This is an invitation for trouble. Kids are taught that when you are home alone, don't answer the door and if you answer the phone say, "Mom's busy. Can I take a message?" Lindsay told a complete stranger where she was going, with whom and that there would not be an adult around. Not a smart decision.

***Speeddemon: OH YEAH!!!! Been thinking bout u since I saw u at Sat's game***

***Cheercrzygrl: u wer there? knw u wer there 2 games ago ddnt knw u came to last 1***

***Speeddemon: yup we chatted here, you seemed so cool, had to see you again. Have not stopped thinking about you since***

Here is a man "*Speeddemon's*" age (25) thinking non-stop about a young teenaged girl. This indicates danger. "*Speeddemon's*" actions could be classified as part of a grooming process. He wants to make Lindsay feel special and appreciated. Who doesn't want to feel that way? The problem is he doesn't know her and she is falling for what he is saying. "*Speeddemon*" is not speaking the truth; he is saying things that he knows a 13 year girl would want to hear.

The second, and very big, concern is he claims that he showed up again at an event that he knew Lindsay would be attending. The actions that he is taking can lead to dangerous and frightening situations. When someone is being stalked it's a very difficult thing to deal with.

Stalking can come in several forms. Physical stalking like showing up to a football game just because you can't stop thinking about a person is one kind. You can also be stalked on-line. Every time Lindsay logged on, who was there waiting? Here is a thought for you. Do you think Lindsay is the first girl that *"Speeddemon"* acted this way with? Probably not.

***I put in a search for information about Mandela and quickly came up with a page that would work if I needed a cover. I minimized that and logged onto Instant Messaging.***

Once you have to come up with ways to sneak around and do something, it should trigger in your brain, *I probably should not be doing this.* We all make mistakes. We all lose sight of safety at times, but when you plot and plan ways to get away with things you know you shouldn't be doing, trouble is just around the corner. This type of behavior is not a mistake; it is a conscious decision you're making and it is wrong and not safe.

***Speeddemon: can I ask you a question***

***Cheercrzygrl: yup***

***Speeddemon: you got a bf?***

Creepy! Nothing good can come out of a conversation when an older man asks a younger girl about her boyfriends or sex life. NOTHING!

***Speeddemon: oh, yeah, almost forgot LOL you ever make out with a guy***

At this point Lindsay should have sensed something was very wrong and ended this conversation. Just because it may have made her feel older and cool does not make it okay. This type of behavior is completely out of line and should not be happening. Parents, if you think this type of behavior doesn't happen to kids, or can't happen to *your* child, keep in mind that studies show 1 out of 7 kids receive unwanted sexually explicit messages from people on the Internet. If your child has seven friends, could he/she be the one who already received one? This type of behavior happens every day on the Internet.

> ***Speeddemon: and we will not leave theater so don't worry about trouble, so dark no one will know we meet I don't want trouble either.***

If there was nothing wrong with this situation, would the adult even worry about getting into trouble? His comment about not wanting to get into trouble says it all. He knows what they are doing is wrong. No adult should have secrets with a youngster. Secrets like this only lead to bad or tragic situations.

> ***Speeddemon: you give me your cell number, I can text you when I get there, you tell me what movie and I will sit in back, when you say you are going to bathroom, come see me***

We have already talked about giving someone your phone number and how that can provide way too much personal information. We now know that is wrong. The big concern is how fast he came up with this plan. It almost seems like he has been thinking about this or even worse, has done this before. If someone is that quick to suggest such an illegal act, it is probably someone kids should not have a friendship with.

As we have seen, things that happen on the Internet can spin out of control quickly. Parents' number one goal is protecting their children and keeping them safe. Easy job? Not always, especially if we are not 100% sure what our children are doing and they have access to communicate with strangers from all over the world. Do you know if your child is communicating with strangers? If you have never sat and looked over your child's "buddy" lists, you can't attempt to answer that question. Kids will refer to on-line strangers as "buddy" or "friend". If they only know them from the Internet, they are STRANGERS.

Here is a great way that families can prove to one another that they are being safe. Open your instant messaging program - yes both parents and kids. Look over the screen names that are on your lists. For each screen name, if you can give the person's real first and last name and identify where you know them from, then they're safe. If you scratch your head, pause even for one second or say, "Who is that?" to any screen name, remove it. Also, you should make sure that your instant messaging program is set so that only approved people on your list may communicate with you.

Protect yourself on the Internet by guarding your identity and personal information. It's the easiest, and smartest, way to stay safe when using technology.

# ISSC Answer Key

1. FALSE
2. FALSE
3. TRUE
4. FALSE
5. FALSE
6. TRUE
7. FALSE
8. FALSE
9. TRUE
10. TRUE
11. FALSE

# CHAPTER 8

## Bryan's Burden

**Profile:**
12 year old in sixth grade
Lives with grandparents
Grandmother: Advertising Copy Writer
Grandfather: Photographer for local newspaper
No siblings
Town demographics: 18,300 residents

Some ideas are better left as ideas. As I learned in English class, ideas are nouns. Most people wouldn't care what part of speech ideas are, and you know what? I don't, either.

English has never been my favorite subject. Who cares if the words in the sentence are verbs or nouns or conjunctions? How often are you talking with someone when they say, *"Hey! I really like that action word you used. What part of speech is that?"* or *"Wow! Awesome adjective, dude."* No, no one does that. Maybe sometimes my grandma does because she works in advertising and says word choice makes a difference, and okay, Mrs. Rickert, my English teacher, but real people don't. In fact, it was because of Mrs. Rickert that I ended up in a huge mess.

Really, it was because of *me* and *my* choices that I'm in this mess, but it started with her.

First you need to know about my situation. I've lived with my grandparents since I was six years old. I moved in with them after my mom and dad died in a boating accident. I was on the boat, too, but I was wearing a life jacket, so I was rescued. Sometimes I think about how life would be different if my parents had survived. Or I think about the fact that I could have died, too, but since I can't change the past, I try not to think about it too much.

For the most part, my grandparents are cool, except when it comes to discipline. They seem a lot stricter than regular parents. I guess it's because they've already been through this raising kids thing, so they know what works and what doesn't, which is unfortunate for me.

My grandparents are fair most times. If I want to do a sport, like play soccer or lacrosse, they support me and take me to practices and come to my games when they could be doing things they want to do for themselves. Because of this, I also don't want to disappoint them, but sometimes I do some dumb things. The situation with the English project was a good example, and to me, Mrs. Rickert went way overboard with her reaction. And I guess my reaction to *her* reaction was way overboard, too.

On that day, she gave this lame group project to make a poster using the different parts of speech. On the poster we had to include pictures that went along with the words we chose. I was in a group with Tim, Ella and Nikki. I liked that group because Tim and I hung out together and played basketball sometimes at recess, and Nikki was okay, too, but to get in a group with Ella was perfect. I've liked her since fourth grade, and I'm not sure, but sometimes I think she might want to go out with me. Since we were working in the same group I thought maybe it would give me a chance to ask her out.

Ella and Nikki came up with the idea for the poster design. It would be set up with little pockets for the words so we could change them every day and still keep the old ones. I was all for "acing" this project, so Tim and I went along with what they wanted. Since Ella and Nikki are always good workers, Mrs. Rickert left our group alone in the corner. I liked that, too, because then we could have fun while we were doing this.

Our group decided that we would use construction paper and cut things out for the pictures. Our plan was to make them into shapes that could be moved and changed to make other pictures which would symbolize other nouns.

For two days we worked hard on this. For example, we cut out a long thin triangle and a couple of circles. With those shapes we could make an ice cream cone on one day, another day a face with two eyes and a long nose, a snowman another day, and things like that. It gave us the opportunity to constantly change the words, which I'm sure Mrs. Rickert would love. I had to give the girls props; Tim and I probably wouldn't have come up with something that cool.

The trouble started when we had to decide how to use the shapes. We'll just say that after a while my creativity kicked in. I looked around to make sure Mrs. Rickert wasn't near us then had a little fun. It was all me; no one else in my group touched the shapes, but their amusement fueled my creativity. The girls giggled at some of the things – I mean nouns – I was making out of the shapes. I know some of them might not have been appropriate, but it's not like I was going to put them on the poster. I got so into it that I didn't notice Mrs. Rickert come up behind me just as I finished setting up the shapes in the most inappropriate ways.

Because I was looking down and concentrating on getting a good laugh out of Ella, I was annoyed when someone tapped me on the shoulder.

"Just a sec, I'm not done," I said, thinking it was Tim. I moved two circles into a shape and was ready to move the rectangle when Mrs. Rickert cleared her throat behind me. I must have shot ten feet straight up in the air. I wouldn't have been surprised to look down and see my skin laying in a heap on the floor because I jumped out of it.

"Could I see you in the hallway, Mr. Taylor?"

Uh-oh! Calling me Mr. Taylor was not a sign of good things to come. Until we stepped into the hall, I hadn't seen Mrs. Rickert's face. Once she turned around I was shocked at how red it was and how dark her eyes could look. Wow! I'd never seen her so angry, at least, not at me.

"Would you like to explain yourself?"

Could I answer, '*No, I wouldn't like to. Thanks, anyway,*' and get away with it? I doubted it.

"I was goofing around," I said. My voice sounded a little squeaky, which was embarrassing. "We were all done with the project."

Mrs. Rickert crossed her arms over her chest. "And what will your grandparents say when I tell them how you chose to use your group's time?"

I gulped. "You're calling my grandparents?"

"Yes, it's my policy when I give a detention."

My heart sank. Just last night I had convinced Grandma and Grandpa that I was old enough for a cell phone. My grades were good and I was doing all of my chores without being asked. They said they would discuss it and give it serious consideration. I thought I'd presented a good case, but if Mrs. Rickert called home that cell phone idea was history.

"I'm sorry," I managed to croak.

"I'm sure you are, Mr. Taylor. Your detention slip will be in the office at the end of the day."

Mrs. Rickert pivoted and opened the door, holding it for me so I could go in first. Of course all eyes were on me when we came back in. I crossed the room and returned to my group. Tim, Ella and Nikki had already rearranged all the shapes into appropriate nouns. All we had to do was attach them to the poster, but I didn't care. At this point, getting an "F" on the project wouldn't have been any worse.

After school my best friend, Seth, already had a seat on the bus when I flung my backpack under the seat and plopped down on the vinyl next to him.

He punched me in the shoulder. "Dude, I heard what happened with Rickert fifth period. Detention, huh?"

"Yeah. I'm dead." I slid down and mashed my knees against the back of the seat in front of us. "So much for the cell phone."

"Aw, sorry, man." He pushed himself into the corner by the window.

"Rickert's a jerk. I swear she's out to get me. Last week my locker got stuck so I got to class a minute late and she gave me a recess detention."

Shannon Archer, a girl in my homeroom and English class, poked her head over the top of the seat in front of us. "You were more than a minute late, Bryan, and it was the third time in a week."

Seth popped forward coming face to face with her. "Butt out, Archer! Nobody asked for your two cents."

She scrunched her face into a know-it-all expression, and you could tell she wanted to stick out her tongue, but instead she slithered back out of sight. "Morons," she muttered.

The name-calling didn't faze me. All I could think about was how close I was to having my own cell phone. I'd even gone on-line to research exactly which one I wanted.

Seth slid down next to me. "Whatcha gonna do?"

"What do you mean what am I going to do? I'll probably be grounded for a month and who knows when I'll get that cell phone. Probably never!" Suddenly I realized my hands hurt because my knuckles were clenched so tight in my lap.

"That's not what I mean." Seth glanced at the people around us then lowered his voice. "You've gotta get back at her, man. You can't let her get away with this."

"Huh? Who?" I wondered if I looked as confused as I felt.

"Rickert," Seth clarified.

"What am I supposed to do?"

He glanced around then leaned in a little closer. "Scare her."

My stomach was already in knots from frustration. "What are you talking about?"

When I looked at Seth he had that grin on his face that meant he was thinking something evil. "Do something to her car. You know, cut the tires or something. Put something really bad in her desk drawer."

My voice got kind of loud. "What? Are you nuts? You know how easy it would be to get caught?"

"I'm just tellin' ya, if she won't leave you alone it's time to get her back before she really gets you in trouble. You don't wanna slash her tires then leave threatening notes on her desk or something. Tell her she better watch her back."

Although the idea of getting Mrs. Rickert back was appealing, Seth's ideas were kind of crazy. If I left notes on her desk someone would catch me coming or going for sure. I couldn't do anything where someone could possibly see me and catch me in the act.

As soon as that thought came into my head, so did the answer to what I could do. I'd wait and see what my grandparents said, but if the cell phone was history then Mrs. Rickert was going to be sorry she did this to me.

My grandfather's gray Honda was parked in the driveway when the bus pulled up. That wasn't unusual because, as a photographer for the local newspaper, he was on his own as far as when he came and went. He got the job done, so that's all his boss cared. And he was really good at it. His office was lined with plaques and certificates from awards he'd won over the years.

When I went in the house I had to walk past Grandpa's office. He had his back to the door while he worked at the computer. It looked like he was going through photos he'd taken. I quietly set my backpack on the bench in the hallway. The floor creaked as I stepped past his door.

"Good day at school, Bryan?" He never turned around. My grandparents are always calm, even when they're really angry, so I didn't know if Mrs. Rickert had called them or not.

I stopped and stood in the doorway. "Uh, it was okay, I guess. How was your day?"

"Fine." He swiveled his chair to face me. "Until about an hour ago when your English teacher called." His eyebrows raised in that classic *What do you have to say for yourself?* expression.

My shoulders drooped and I sagged against the doorway. "You don't have to say anything more."

"No, I don't. And I'm sure you'll understand why that cell phone conversation is now put on hold."

I nodded without looking up at him, wondering if his choice of words was supposed to be funny. I wondered just how long "on hold" would be.

He swiveled back to his desk and resumed working on the photographs. "Have a snack then get your homework done before your grandmother gets home," he said over his shoulder.

"Yep." I picked up my backpack and skulked off to the dining room. At least I hadn't been banned from my one hour computer time after dinner. Or, I didn't think I was. I supposed anything could change once Grandma got home.

I had trouble concentrating on my homework because all I could think about was how one stupid phone call from a jerky teacher messed up my chances of getting my own cell phone. Out of all my friends, I was the only one who still didn't have one. It made me feel like a baby when they'd talk about their text messages, or tell someone to text them. The more I thought about what I was missing out on, the more frustrated I became. And the more frustrated I became, the more I thought about what Seth had said on the bus.

*"You've gotta get back at her, man. You can't let her get away with this."*

He was right. She didn't need to ruin any other kids' lives the way she was ruining mine. It was time to make her stop and think before she handed out detentions like she thought they were candy.

At dinner Grandma gave me the "we're disappointed in your behavior" lecture and she supported Grandpa on the cell phone issue.

So, it was a done deal. I had a plan. I was following through on it.

The computer was on the desk in the corner of the living room where my grandparents could monitor. They were good about giving me privacy, even with it being right out in the open. After dinner Grandma sat down to watch *Wheel of Fortune* and knit and Grandpa read a photography magazine. I chatted on-line with a couple of friends at first. Spring training for professional baseball was just starting so we were debating the best batters and pitchers. But I couldn't get Mrs. Rickert off my mind. I knew how I was going to get back at her.

I went to my school's website and clicked on Mrs. Rickert's school e-mail. I didn't write anything in the subject line, but in the message I wrote:

***U R THE WORST TEACHER EVER***

Of course I wasn't going to sign my name because that would be really dumb. I used the mouse to move the arrow over send and was ready to click it when I realized how dumb even *that* was. The message would be coming from my e-mail address. It wouldn't take her long to figure out I'd sent it, even without me signing it.

I sat back and stared at the screen. How could I do this without getting caught? Then it hit me. I'd create a new free e-mail account that no one would know about except me. I wouldn't even tell Seth. Score! I had my answer.

I clicked on the browser then opened up the bar for creating e-mail accounts. I couldn't believe how easy it was. I made up a fake name: Stuart Little. Creating my fake account using an animated mouse's name made me laugh out loud, causing my grandparents to turn and look at me.

"Sorry," I said. "Just got a funny message from one of my friends." Of course, it wasn't the truth, but it wasn't like that fib was going to hurt anyone. I continued filling out the information for the account, making up all of it so there was no way for anyone to know it was me. I made sure the birthdate made me 18 and wrote that I lived in Australia. I smiled at my cleverness. I'd always wanted to go there, so why not let my made up person live there?

Next I had to have a screen name or ID and a password. The e-mail program gave me some suggestions for screen names, but I knew exactly what I wanted. For screen name selection I typed in *TeacherExterminator* and I made the password *IhateRickert2.*

Just setting up this account made me feel better. No one in the school liked her, so I was doing everyone a favor. She'd be sorry that she's such a jerk. The last thing I had to do to create the account was to set up the answer to a secret question in case I forgot the password. I chose the question *What is the name of the town where you were born?* I typed in *Middletown* then clicked *Create Account.*

Done.

I clicked on compose mail and typed in Mrs. Rickert's school address. In the subject line I put *IMPORTANT!!!!* That would get her attention. Now that I had the e-mail open I wasn't sure what I should write. The way she walked around school, she acted like she was the queen. Maybe it was time for her to know the truth. I put the mouse in the text box and decided to send a simple message.

***U stink! u r a horrible teacher***

I thought about signing it Stuart Little, which made me laugh, but changed my mind. I wanted her to take this seriously because she had seriously messed with me. For just a moment I hesitated with the mouse hovering over the "send" button. Once I sent the message there was no way I could take it back. Finally, I decided it didn't matter. She hurt me; I was hurting her. I clicked "send" and received the message that my mail had been sent. I couldn't wait until tomorrow.

To say the results were disappointing was an understatement. I'm not sure what I expected when I got to class, but I thought Mrs. Rickert would at least look upset. Instead, she acted like nothing was different. Had she even looked at her e-mail? All day I kept waiting to hear something from someone, but nothing was different. Mrs. Rickert even laughed and smiled when she talked to other teachers in the hall. That really bugged me. I guess Seth was right; someone needed to scare her.

It was easy with the fake e-mail account to send whatever I wanted because I knew she wouldn't know who sent it. That afternoon, as soon as I got home from school, I decided to send another message. Mrs. Rickert always parked in exactly the same spot in the parking lot. I noticed lots of teachers did that, even some of my favorite ones. Since she didn't seem to care that everyone thought she was a horrible teacher, I thought maybe she'd care if she thought someone was going to do something to her car. I didn't put anything in the subject line this time because I figured she'd recognize the *Teacher Exterminator* e-mail address and open it.

That message said:

**u should be nicer. I know where you park ur car.**

After sending that one, I was sure when the bus pulled into the school parking lot the next morning that Mrs. Rickert's car would probably be parked somewhere else. But it was in its usual spot. I figured maybe she hadn't received that e-mail yet, either, which frustrated me. They weren't bouncing back, so where were they going?

Finally, the following day there was confirmation that I was successful. When the bus rolled past the lot, I craned my neck over Seth to look. Not only was her car not in its usual spot, after a quick scan, I could see it wasn't in the lot at all. I sat back in the bus seat and smiled. I had power over Mrs. Rickert. I thought of it as revenge for everyone else who hated her, too.

Of course the threats didn't mean anything because, like I told Seth, I wasn't going to slash her tires or leave notes on her desk where someone could see me doing it. But I had to admit it was a cool to know I'd finally scared her and it was all done anonymously. Well, as anonymous as *Teacher Exterminator* could be.

I waited a couple of more days before sending the last message.

***u better watch ur back***

I wasn't sure if it was my imagination or not, but after that it seemed like Mrs. Rickert *was* nicer. It didn't help me get my cell phone any sooner, but I felt like Superman. Maybe I'd helped make things better for all the rest of the kids at school, too.

A month later I again started dropping hints about the cell phone to my grandparents. Everything had been cool at school, and I thought they might have even forgotten about the poster incident in Mrs. Rickert's class. Mrs. Rickert was also parking in her regular spot again. Life was back to normal.

One time in class when we were reading a story together, I glanced up and thought I caught Mrs. Rickert staring at me with a funny look on her face. Another day, the guidance counselor called me down during study hall to ask how things were going. She said she would be calling everyone down at some point, but I still hadn't seen anyone else go. I thought maybe it was because of my parents being dead or something, but it still seemed random.

All of the clues I'd missed came together one night just before dinner. I was finishing my homework in the den when the doorbell rang. Grandpa answered it, and I couldn't see who was out there, but I saw him step out onto the porch.

"Who's at the door?" Grandma called to me from the kitchen.

"I don't know. Grandpa answered it." I hurried into the living room and peeked out the window. A police car was parked out front, and then I recognized the school resource officer and another policeman standing at the bottom step. I wondered why Officer Holbrook would be at our house. A few minutes later I had my answer.

Grandpa brought them into the house and led them to the living room to the computer. "Bryan uses this one," he said. My knees felt like jello. Why did they care what computer I used?

Grandma stepped to the doorway and was drying her hands on a dishtowel. Her forehead wrinkled like it always did when she was worried.

"Have a seat," Officer Holbrook said to me, pointing to the chair in front of the keyboard. I sat in the chair and looked up at him. "Now, can you log into your e-mail?" he asked.

My fingers shook, but I typed in my screen name, *Bballstar44* and my password. As soon as the screen came up, the instant message box popped open with a couple of messages. One was from Seth. I stared at them. This wasn't the time to get into a conversation. I glanced at my inbox, but everything looked fine there. I let out the breath I'd been holding.

"How about the other e-mail account?" Officer Holbrook asked.

My heart tripped. "Wh-what other e-mail account?"

"The one you used to send Mrs. Rickert threatening messages," Grandpa interjected. "*Teacher Exterminator* ?" Even though it sounded like a question, it was obvious Grandpa knew.

It was no use playing dumb. I logged into my secret e-mail account then dropped my hands into my lap. All of a sudden having a cell phone didn't seem so important anymore. I was more concerned about whether sixth graders could be arrested.

# I.S.S.C.
# Internet Safety Savviness Challenge

How much do you know about on-line harassment? Test yourself with these questions, answering true or false. Check your answers with the answer key at the end of the chapter.

1. ____ The school can take enforcement action over what Bryan did.

2. ____ What Bryan said in the e-mail breaks no laws.

3. ____ The teacher is partly responsible.

4. ____ The grandparents, as guardians, can be held liable for Bryan's actions.

5. ____ It's possible to set up an e-mail account that can never be traced.

6. ____ Because of the age restriction of 18 to open an account, Bryan broke the law when he lied about his age.

7. ____ Bryan can avoid consequences for threatening the teacher by saying he didn't mean to scare her and he was only joking.

8. ____ Law enforcement and Internet companies routinely check the activity on everyone's account to make sure they're not breaking the law.

9. ____ Even though the teacher punished Bryan for his inappropriate behavior in class which caused him to be upset and send her the messages, Bryan is still the only one responsible for the outcome of this situation.

10. ___ After hitting "send" on an e-mail, you have fifteen minutes in which you can go back and "recall" that message so it doesn't go to the recipient's e-mail in-box.

# What Can We Learn?

Bryan made some choices that he will have to live with for a long time. These choices also could have very serious consequences. When we revisit Bryan's actions and thought process, there were some warning signs that should have told Bryan *DON'T DO THIS*!

> ***I realized how dumb even that was. The message would be coming from my e-mail address. It wouldn't take her long to figure out I'd sent it, even without me signing it.***

Right here is where Bryan should have stopped and reconsidered his plan. If you have to worry about someone finding out if you sent a message, that message should not be sent. Common sense tells us that this is not a good situation. Bryan let his emotions get the better of him, and he did something that he knew was not right. Bryan knew it was wrong, yet he did it anyway.

When Bryan created the "new free e-mail" he thought he was being sneaky and would get away with threatening and harassing Mrs. Rickert. As we know now, that is not the case at all. When you connect to the Internet using your computer, several things happen. First your computer connects to your Internet Service Provider (ISP). The ISP is the company that provides your home or mobile device access to the Internet. Without an ISP you can't connect to the Internet.

When you "log on" with your computer and go to the Internet the ISP assigns an Internet Protocol (IP) address. The best way to understand IP address is to think of it like a phone number. You can have phones in your house, but if you do not subscribe to a phone company, the phones will not work. When you subscribe to a phone company, they assign your house a phone number and then turn on the service. Now, when you go to use the phone, you are able. When you call someone and they have caller identification, they know it is you because your phone number shows up. The phone company will keep records on your account, name, address, billing information and other important documentation. They can also, when requested, provide logs showing incoming and outgoing calls and other important information.

ISPs can do the same thing and provide the same information. When Bryan signed up for his free e-mail account his IP address was logged. When he sent the messages to Mrs. Rickert his IP address was logged. All Officer Holbrook had to do to identify "Teacher Exterminator" was locate the IP address in the e-mail. Once Officer Holbrook found this information, he would apply for court orders to identify what ISP owns that IP Address. When the ISP is identified, another court order to the providing company will tell them who had that IP address assigned to them at the date and time the message was sent.

These steps are why it took approximatley a month for Officer Holbrook to arrive at Bryan's house. Although it may take some time to work through all the paperwork and court procedures, the information will be there and will be found. Bryan thought he was being sneaky and was never going to get caught. Ten years ago that may have worked, but not today.

Even on a smaller scale of safety, when Bryan signed up for the e-mail account, he stated that his age was 18. Kids should understand, if they have to lie about their age to do something on the Internet, there is a reason they're being restricted. The Internet offers several situations (chat rooms, web sites, auction sites), both good and bad, that are not appropriate for children to be involved with. If there is an age requirement, kids must respect the restriction because it is in place to keep kids safe.

***For just a moment I hesitated with the mouse hovering over the send button. Once I sent the message there was no way I could take it back.***

Throughout this book there's a message that has been reinforced. If you pause because you're thinking what you are about to do is wrong or not safe, STOP! Bryan's pause tells us he knows what he's doing, he is just not considering all of the possible consequences of his decision. His only correct thought was that he could not take the e-mail back. Once that "send" button is pushed you have no control over that message and can't stop it. It is gone from your computer but not gone from the Internet. Bryan made some decisions out of anger and frustration. If he had followed his first instincts, he would probably be talking on his new cell phone right now.

***Teacher Exterminator***
***u stink! u r a horrible teacher***
***u should be nicer. I know where you park ur car***
***u better watch ur back***

Every state has different laws, but almost every state has laws against stalking, threatening and harassing. Do these comments and this behavior apply? In most states, I would say yes. No one has the right to make another individual live in fear or be constantly bothered. The comments that Bryan made were meant to make Mrs. Rickert afraid. That is against the law.

Here is where the Internet allows us all to cross lines that we would never cross in the "real world". There's no way Bryan would walk up to Mrs. Rickert and tell her he changed his name to "*Teacher Exterminator*" and she should "watch her back". Bryan would have been removed from the school immediately, suspended and would face criminal charges. But on the Internet he felt brave because he thought no one would know it was him.

A note to parents: Bryan's grandparents had a great safety rule in place. The computer that Bryan used was located in a "public" area of the home. The unfortunate aspect is it seems they did not check up on Bryan while he was on-line. Keeping the computer in a common area of the home is a great safety plan, but if you are not checking in from time to time or can't see the computer screen, this may provide the children a sense of security that they can explore or push the limits knowing they are not being monitored.

Anything you do on the Internet can be traced back to you. For everyday use this isn't a concern, and there would be no cause for authorities to check your Internet activity; however, if you engage in inappropriate or illegal activities while using the Internet the potential for involvement by law enforcement exists. An easy general rule to remember is, if you would not say something to someone's face, or you would not send it as yourself, then don't say it over the Internet, either.

Finally, threatening and harassment are crimes; it doesn't matter if you say it in person or use technology, a threat is a threat and harassment is harassment. In the scenario presented, if Bryan had taken responsibility for his actions in class instead of blaming Mrs. Rickert, he would have never put this burden on himself.

# ISSC Answer Key

1. TRUE
2. FALSE
3. FALSE
4. TRUE
5. FALSE
6. FALSE – The age restriction is suggested but is not enforceable by law agencies.
7. FALSE
8. FALSE
9. TRUE
10. FALSE

# CHAPTER 9

## Christine's Confusion

**Profile:**
14 year old freshman
Interests: Theater and Music
Mother: Pharmacist
Father: Accountant
Siblings: 12 year old brother, Robbie
Town demographics - urban: 217,000 residents

When you have a best friend, that friend will have your back no matter what; at least that's what we expect, right? Sherry has been my best friend since pre-school, and there have definitely been times when we've had to stick up for each other. We've only had a few minor fights over the years and usually it was because we both wanted the same thing: a boy, the first seat in orchestra, or the same dress for the Spring Fling Dance. In the past our fights only lasted a few hours and, at the most, a day or two, but this last time our disagreement went too far.

The problem started over the roles for the school's spring musical, *Fiddler on the Roof.* Even though I'm only a freshman, I was hoping to score one of the sister roles. It was pretty clear which upperclassmen were going to get most of them, but everyone was telling me that the last one was between me and Tracy Weston, a sophomore. The unspoken rule was that freshman never got lead roles in the senior high musicals because they had to "pay their dues and wait their turn". But everybody also knew that I'd been getting the lead roles in plays since elementary and middle school, including shows with the local theater group. With my history, I didn't think Tracy had a chance.

Because it was the night when I was waiting for Ms. Reynolds, the drama director, to post the roles on her website, it's easy to recall exactly when the latest problem between Sherry and me started.

The soundtrack for *Fiddler* blared from my i-pod speakers. I sat on my bed singing at the top of my lungs and painting my fingernails a bright shade of green for St. Patrick's Day when there was a knock on my bedroom door.

I kept my focus on my big toe as I spread the polish in upward sweeps. "Ye-e-e-ss?" I sang the greeting as if it were part of the song.

The door swung open and Sherry stepped in. "Hey, your mom said to just come up."

"I wondered when you were going to get here. Ms. Reynolds said the list would be posted by 7:00." At the same time we glanced at the clock on my nightstand. "Half an hour." I returned my attention to my toes and picked up singing *Matchmaker*, imagining myself on stage in the role.

Sherry plopped into the chair at my desk and draped her arms over the back with her chin on her forearms. "Aren't you nervous? I mean, I'm sure you did great, but everyone is saying the role of Hodel is between you and Tracy Weston. Did you see her audition?"

"No, they wouldn't let us watch the others, but she can't sing."

"How do you know if they didn't let you in her audition?"

I rolled my eyes and looked up at Sherry. "She was stupid enough to post a video on her *FriendPost* profile."

Sherry's eyes opened wide. "You're friends with her?"

"Only on-line. There are a bunch of Drama Club people on-line, so we've *friended* each other." I dipped the brush in the bottle one more time. "I'd never be friends with her otherwise. She's so weird."

"Can I see the video? I don't even know what she looks like."

"Yeah, sign in to my account and go to her page. It'll be under her videos on her page."

Sherry swiveled on the chair and reached for the mouse. While I finished painting my nails on my left hand, her fingers tapped away on my keyboard.

She stood. "Okay, do you want to come over and put in your password?"

I held my hands with the wet fingernail polish up toward her. "Does it look like I can type in a password? I'm not messing up the polish over Tracy Weston. I'll just tell you what it is."

Sherry sat back down. "Are you sure you want to give me your password?"

"You're my best friend. Why not?" I blew on my fingernails then said, "It's Broadway Diva."

Sherry laughed. "Broadway Diva? Really?"

"I'm going there some day," I said. I'd had that dream since my mother and grandmother took me to see *Beauty and the Beast* on Broadway when I was six years old.

"If you want to go now, they have tickets available for any show you want to see." She put in the password.

I glanced up but she was facing the computer screen. "Ha! Ha! Very funny," I said. "I *am* going to be a star some day."

"Well, you aren't called the drama queen for nothing."

I started to protest, but Sherry interrupted. "Okay, I'm on her profile." The screen changed as she clicked through pages. "Here it is." A couple of clicks later and Tracy Weston was filling the screen, a microphone just inches from her mouth.

I flapped my hands to dry my nails as her singing started. The song was *Somewhere Over the Rainbow*. "She looks like she's trying to eat the microphone. She's such an amateur."

"She sounds pretty good, though," Sherry said.

"You think that's good? She's off key and she's just so…so blah. All she does is hold the microphone. There's no expression on her face. She's a trashy zombie."

Sherry glanced over her shoulder at me. "That's kind of harsh isn't it?"

"You think she's any kind of competition for me?"

"She's pretty good, actually. I think you might have some competition." Sherry turned back to the screen. I got up from my bed and crossed to the desk just as Tracy warbled the bluebird part.

"I thought we were best friends? Where's your loyalty?" I asked.

"This has nothing to do with loyalty," she responded. "This is all about honesty. Tracy's just as good as you are."

"Are you kidding? She's a loser." I hip bumped Sherry to get her out of my chair then sat down to pause the video. There was a space for comments underneath the video, so I clicked on it and typed.

> ***Somewhere over the rainbow? Yeah, that's where this singer belongs. I'll show you how this song should be sung.***

"What are you doing?" Sherry asked as she leaned in to read my comment.

"What's it look like?"

"You're going to write something mean like that? Don't you care what others will think?"

"They'll read this and agree. She doesn't deserve a lead role."

Sherry grabbed at the mouse under my hand, but I pulled it away. "They'll look at that and think you're horrible and conceited," she said

"No, they won't. She's not even popular." Sherry tipped her head at me in obvious disapproval. I sighed dramatically. "Okay, fine." I deleted the comment and clicked out of Tracy's page. "Who do you think people care more about, her or me?"

Sherry had a strange look on her face. "Wow! I think the possibility of stardom is going to your head. I've never heard you talk like this."

I clicked out of that site and went to Ms. Reynolds's site to see if the cast list was posted. "Being in high school is different than middle school, Sherry. This is where popularity really counts. You want people to listen to what you have to say."

"So, say something nice."

"There isn't anything nice to say. I am clearly the one who should play Hodel." I studied my fingernails to make sure I hadn't messed up the polish by typing before it was dry.

"I'm surprised your password's not Broadway Snob."

"Hey, you can thank me for you being so popular at school." I clicked through Ms. Reynolds's site until I found Drama Club.

"What do you mean?"

"Do you think anyone else would even notice you if it weren't for the fact that we're best friends? I totally carry you at school. You're popular because you hang with me."

Sherry frowned and started toward the door. "Really? Is that what you think? I'm nothing without you?"

I didn't understand why she was getting so worked up. She was the one who started this discussion. "Would Douglas Perry talk to you if I wasn't around?"

"The world doesn't start and stop with you, Christine." The tone of Sherry's voice rose, a sure sign this was going to be one of our big fights.

"I didn't say it did." I turned back to the computer. "Come on. Let's see if I got the role."

"I don't care if you did. I think Tracy Weston is just as good as you are, and you're only a freshman. She should get the role."

What Sherry said didn't totally register in my brain because I was distracted by the website. "Oh, here it is. Ms. Reynolds posted it." I scrolled down the list but didn't have to go far to find my name. I jumped from my chair and danced around. "I got it! I got it! I knew it was all me. Weep into your cereal over that one, Tracy Weston. I owned you."

The straight line across Sherry's lips never wavered.

Unable to contain my excitement, I grabbed Sherry's arms and shook them. "Aren't you excited for me? I got what I wanted."

"It always seems to work out that way, doesn't it, Broadway Diva." There was a sneer in her voice when she repeated my password, and the distance in it went along with her actions when she stepped away and reached for the door handle. "I always thought you were just confident. I had no idea you were actually so conceited." She swung the door open. "Enjoy your success. It's going to be lonely celebrating alone." She slammed the door and was gone.

For a minute I stared at the poster of my favorite rock band on the back of my door. Who was she kidding? That was the fun of being popular. Everybody would be excited for me. I shrugged off the argument and went back to the computer. Sherry and I had been friends forever. This would all be over by lunch tomorrow. All I had to do was give her my sad face, tell her I was sorry and it would be just another tiff forgotten.

I ran downstairs to tell my parents that I'd gotten the role. They were just as excited as I was. My father even started singing a few lines from *If I Were a Rich Man*, the song Hodel's father sings in the musical. Their reaction was a lot better than Sherry's. I figured by tomorrow, when she saw how excited the rest of our friends were, that she would be excited for me, too.

I went back to my bedroom, logged onto the Internet and went to my personal page. I'd barely updated my status when friends started sending messages to congratulate me. This was definitely one of the best nights of my life. I was surprised when a message came through from Tracy Weston. She simply wrote, "Congratulations on getting Hodel." I started to respond then decided to not say anything. "Sorry" seemed kind of trite, so I left it alone.

Curious, I clicked onto her profile. In her status she had simply written the message, "It's time to make lemonade out of the lemons." Several of her friends had already posted on her profile. They wrote messages like "chin up", "u will do gr8" and "sorry you didn't get

Hodel". I was glad she had supportive friends, too. I switched back over to my profile and read more of the messages that I was receiving. I felt like a star and could hardly sleep when I finally went to bed a few hours later.

The next morning started out like every school day. I showered, dressed, ate breakfast and checked my e-mail and *FriendPost* page before leaving for school. My inbox was full of e-mails, so many that I probably wouldn't be able to read all of them before school, but I decided to read a few. I figured my friends were just excited for me and wanted to congratulate me.

The first couple of e-mails were exactly what I thought, but then I opened one from a girl who was a friend through Drama Club, and I was confused.

***Jenna Torbett: How could you be so mean?***

The e-mails from other friends after that one weren't any better and didn't make sense.

***Randy Newcomb: Wow! That was low. I was excited 4 u until I read ur comments about Tracy.***

***Kati Stone: I never heard of a sore winner b 4. Why wuld u do that 2 a nice kid like Tracy?***

***Nicole Davidson: i was excited that u got the lead role until i saw what u wrote on tracys wall***

I had no idea what their e-mails were talking about. Even though I had briefly considered it, I hadn't written anything on Tracy's wall. I typed in her name and went to her profile. I was shocked to see several messages about me, but they weren't good messages. They were comments about how nobody loves me more than I do and everyone should just ignore me.

My mouth went so dry I could barely swallow. The words were wavy in front of me, but I was so shocked that I swiped at my tears and kept reading. It seemed like half the kids at the high school thought I was a heartless creep, and I didn't even know why until I suddenly came across a posting with my name on it.

***Christine Zoftlin: If u want to know why Tracy didn't get the role of Hodel go check out her video. Pigs being slaughtered sound better.***

Farther down there was another comment posted about 11:15 p.m. the night before – *after* I had already gone to bed.

***Christine Zoftlin: sorry u didn't get the role of Hodel but we knew I would get it anyway.***

Another comment from my account followed at 11:35 p.m.

***Christine Zoftlin: I hear Broadway calling my name- it's choking on yours.***

I stared in disbelief. Now I understood why I'd received those e-mails from my friends and why Tracy's friends were posting such horrible comments about me. There was my picture and my name next to those nasty comments, but I hadn't written them. I wasn't even on-line when they were posted. I shook my head, bewildered, until it hit me.

I hadn't been on-line, but I knew who had.

I rummaged through the side pocket of my backpack until I found my cell phone. I clicked on Sherry's name in my contact list then hit text message. My fingers flew across the tiny keypad.

***what did u do? thought i could trust u w/ my password. ur sposed 2 b my bf***

I pressed "okay" to send the message then tapped my green fingernails against my desktop. Every nerve in my body quivered, making it hard to sit still while I waited to see if Sherry would respond. I looked back at Tracy's profile. I couldn't even imagine what the kids at school thought of me right now. More people were posting comments against me as I waited for Sherry to respond. Finally my phone beeped three times with the message.

***had to knock u off ur pedestal b 4 u got any worse***

Stunned, I hit reply.

***No one will believe it wasn't me! im ruined at school.***

A text came back right away.

***u wanted 2 b the drama queen now the drama is about u***

The cell phone lay in my palm, and my fingers were paralyzed, as numb as my brain. I never dreamed something like this would ever happen to me. The time on the phone changed to 7:04. Time to walk to school. My stomach somersaulted just thinking about facing the other kids. For a minute I thought about faking a bad headache so I could stay home, but I knew my parents would never buy it. I was fine twenty minutes ago.

I signed off line and grabbed my backpack to go and face the day.

My fears were realized as soon as I stepped onto the school grounds. It seemed like all around the front of the school the kids were clumped together in tight little huddles. I could imagine what they were saying. I knew not everyone would know about the comments on Tracy's site, but word spreads fast in school, so it wouldn't take long.

I looked by the flagpole for my usual group of friends. Relief washed over me when I saw that at least that much was normal. They were all gathering there. When I passed other students, several of them glared at me or quietly called me bad names. Twenty four hours ago they were all wishing me luck at auditions. Today I felt like the scourge of the school.

The confidence that I'd always felt abandoned me like some of my so-called friends. I approached the kids at the flag pole. One of them said something I couldn't hear and walked away. Two just stared at me.

Tammy, a friend since elementary school, stepped toward me with her face just inches from mine. "You know, I always knew you had an ego and only cared about yourself, but WOW! I never realized it was this bad. Pretty cold."

She turned and walked away, and the other girls followed. I wanted to scream and tell the world I hadn't posted those messages on-line, but my emotions were so close to the surface, I knew if I said anything I would start crying.

I weaved past the groups of kids, wishing I could be invisible. When I reached the stairs leading to the front doors, I heard Sherry call me. I pivoted, ready to tell her off, but she had a posse with her, and her evil smirk stopped me.

"Did you hear the news?" she asked. "In your honor, they've changed the musical for this year." I suddenly became aware that the four other people with her were humming quietly. I recognized the song immediately; it was from the Wizard of Oz.

Sherry smiled, but it wasn't a warm smile. "Ms. Reynolds has already assigned your role because there's a song that would be perfect for you." As if on cue, the other four sang, "If I only had a heart" then burst out laughing. I stood frozen in place as they turned and disappeared through the front doors.

There was an ache in my stomach that felt like I'd been punched. Never in my life had I felt so alone and isolated. I wondered if there was any way I'd ever be able to fix the damage Sherry had done using my password. Suddenly the lead role in the musical didn't seem so important.

# I.S.S.C.
# Internet Safety Savviness Challenge

Below are examples of passwords someone might choose to use. On the line after each one, write how the password could be improved to make it safer. Then after reading the *What Can We Learn?* section of this chapter, revise the password one more time using the knowledge you gained about how to create the safest passwords. (There is no answer key at the end of this chapter because answers will vary.)

**1. A computer connected word**
password ______________________________

**2. A mix of letters and numbers**
123abc ______________________________

**3. A pet's name**
Fluffy ______________________________

**4. A school mascot name and sports number**
Spartans30 ______________________________

**5. A home address**
114 Hart St______________________________

**6. A favorite activity**
soccer ______________________________

**7. A birth date**
8181997______________________________

# What can we learn?

Sometimes our friends can be the best thing that ever happened to us. They are there when we need to talk, can offer advice, will listen if we are having a bad day or will be the first people we want to see when something great in our lives occurs. But if we take that friendship for granted or don't respect how valuable friendship can be, we may have some problems.

The reality during childhood is that friends will fight, and when that happens, tempers flare. We know the person so well that in anger the comments we make are sometimes very hurtful. Why? Because we know what our friends' strengths and weaknesses are, so we may attack those out of anger. Fights between best friends can be traumatic enough, but with technology there can be a permanent record of the fight, as with Christine and Sherry.

There are several lessons that we can learn from both Christine and Sherry. Let's start with Christine. Many kids have done exactly what Christine did. You are hanging out with a friend and give your friend your password to check out something on-line. We want to trust our friends, so sharing a password might not seem like such a big deal. Sherry proved why keeping passwords private is important.

Christine should have done exactly what Sherry wanted her to do in the first place.

> ***"Okay, do you want to come over and put in your password?"***

At that moment, Sherry was trying to do the right thing, but Christine did not want to listen. For whatever reason, laziness, nails drying, too tired, or trusting her friend, Christine made a mistake. She should have gone to the computer and entered her own password. If she had, none of this would have occurred.

Obviously this is not all Christine's fault because it is clearly not. Sherry could have used better judgment and decision-making. As angry as she was at Christine's attitude and comments, this was not the proper way to deal with it. Christine had enough trust to give Sherry her password, allowing her to assume her identity. Was that a good choice? Absolutely not, but what Sherry did with it went too far.

Christine thought this was just another best friend fight that would be over by lunch the next day; however, with a few clicks of a mouse and hurtful words, that friendship could be forever ruined. The comments Sherry made are now permanent. Yes, Christine can erase them from her site and maybe she would get lucky and erase them before people saw them, but probably not.

You know that Tracy has already seen the comments. No doubt, as soon as some of the messages were posted, Tracy's friends were in touch with her. You know that Ms. Reynolds, the school principal, and all the teachers will hear about it. Do you think that some parents will hear about it? I am sure they will, and I am sure that some will call Christine's parents and voice their concerns or disapproval regarding what was written. All of this could have been prevented if the password was never shared.

The only people children should share their password with is their parents. Yes, parents should know their child's password. This is not because we do not trust our children; it helps protect them. If something does happen where safety is a concern, parents should have access to gain valuable information. This does not mean that parents should be logging into their children's accounts every day but it is something that, periodically, parents should do to make sure their children are doing the right things.

Passwords are in place to help us stay safer on the Internet. They protect us from people taking information from us, logging in to our accounts and they help protect our identity. Being able to trust our friends is a very important part of life, and it is a great thing to feel that close to someone, but it does not have to include passwords. A true friend would never ask or pressure someone to give up his or her password, and a true friend would also accept an answer of "no" if you refused the request to share it.

We can't take back things that are posted on the Internet, so we have to maintain control of our lives and information. Protecting your password is the most important factor for staying safe on the Internet.

# CHAPTER 10

# Hannah's Hot Water

**Profile:**
15 year old high school sophomore
Member of school swim team
Mother: Pediatrician
Father: Radio Producer
Siblings: sister, Danielle, 17
Town demographics: 28,000 residents

I ripped open the clear plastic bag containing samples of the new team bathing suits and held one up. My sister, Danielle, was one of the captains of our high school swim team, so she was helping the coach and athletic director pick out the new style. Since I'm also on the team, Danielle asked for my opinion. I thought it was cool that even though I'm two years younger that she valued my opinion.

Our school colors are kelly green and gold, so all of the suits were that combination. I held up the first one. It was mostly green with three diagonal strips of gold stars going across the front. There was something rich about the look. I couldn't wait to try all of them on.

I'd been on the swim team since junior high school. Because I had done so well in competitions, when I was in eighth grade I was moved up to the high school roster. It had required a special waiver from the other coaches in our league, but I was proud of the fact that they all thought I was so good that I should compete at a higher level.

I'd worked hard to get to that point. I was serious about the sport, so my parents hired a special coach who had trained Olympic contenders to work with me. I wasn't sure if I ever wanted to go to that level, but for now I was working my hardest just in case. I loved being on the school swim team. Since we had a good reputation, a lot of kids from our high school came to our meets.

My boyfriend, Todd, was a senior and a starter on the varsity basketball team. We'd been going out for almost a year. Since we both were involved in winter sports, we often had practices, meets or games at the same time, so he rarely saw me swim and I only got to see a few of his games. It was cool going out with an upperclassman, especially a senior. A lot of the other girls were jealous.

I glanced at his senior picture in the frame on my desk. It was hard to believe that a year from now he'd be away at college and that most of the colleges he wanted to go to were out of state. I would miss him *so* much when he was gone. As silly as it seemed, I picked up the picture and kissed it. Sometimes I still couldn't believe he was my boyfriend.

I stripped off my clothes and slipped into the first swim suit. It fit perfectly. I opened my closet door so I could look in the full-length mirror. I stepped in front of it, but my phone beeped with a text message, so I went to check it. It was from Todd.

***Movie later?***

***Have to ask 'rents. Trying on new swim team suits now***

***Ooh, want help***

***Lol U r crazy***

***4 u bet you look hot***

***lol guess that's for u 2 decide***

***text me some pix***

***k***

***make em good***

***lol k***

Since we'd been going out, I'd used the camera on my phone to send Todd lots of pictures of me. Those were always the kind where I'd hold the phone way out away from me and take the pictures. Sometimes they were funny and sometimes serious. Just the weekend before I'd taken a picture of myself holding the medal I'd won in the invitational meet.

I held the phone out and took a couple of pictures, but I had to hold it at a really weird angle to show the suit. Then I thought of the time when I was at my friend's house and she'd sent full length pictures to her boyfriend. She stood in front of a mirror and took a picture of her reflection. That seemed like it would probably work, so I went back over to the closet.

It was hard to figure out which angle was best to use. I took one with a side view, but I wanted him to see the cool design on the front, so I took one of those, too. It took a few tries to figure it out so the picture didn't look bad. Finally, a couple of good ones came out. Since I wasn't sure which one he'd like better, I sent both. It didn't take long for his response.

***What do u think?***

***2 much material***

***huh?***

***thought I'd get better pix than that***

***Like what?***

***more skin***

***lol***

***im serious I luv ur body***

It felt good when Todd said things like that. In middle school it seemed like every girl was developing except me. Finally I had a guy noticing that I wasn't a little girl any more.

***<3***

I couldn't resist giving him the "heart" sign. It felt great to be in love and to be loved, even though my mom and dad always picked on me and said it was puppy love. It felt like the real thing to me. I added to my text to him.

***Maybe youll like another suit better brb***

I tried on a couple of other swim suits and sent him the pictures. They showed a varying degree of skin. One was open up the sides, so I thought he might like that better. I took a picture and sent it to him.

***what do u think?***

***getting closer think sports illus***

His answer didn't surprise me, and it actually made me laugh. I checked the pictures on my phone to see if I had a better one I could send. Unfortunately, they all looked pretty much the same. Even though I pretended I didn't know what kind of picture he was asking for, I knew. I'd never crossed that line with anyone before, but I also didn't want to lose him. I figured just one picture couldn't hurt anything. He'd be the only one to see it since it was going into only his phone. I knew I could trust him to keep it to himself.

I wasn't comfortable doing a real nude picture, but I thought if I did something that would make it look like I didn't have anything on, that would be cool. All of a sudden I had a great idea, and it actually made me laugh out loud. I took my yellow bikini out of my dresser and put that on. I'd tell Todd that was one of our suit choices. I didn't want him to see the bikini, so I had to set it up to look like I wasn't wearing it.

On my bed I had one of those pillows that you could sit up against and it was like the back of a chair. I grabbed that and set it on the floor a few feet away from the full length mirror. My stomach fluttered, but I wasn't sure if it was from apprehension or excitement. There was a little thrill in knowing Todd was waiting.

Next, I put on dangly earrings that Todd once told me made me look sexy. Then with a touch of make-up, I was ready to send him the picture he wanted. Well, at least the only one I was comfortable sending.

I sat down on the floor behind the pillow and pulled my bikini strap down so it couldn't be seen in the reflection. I angled myself so I could drape my bare arm and leg over the back and sides of the pillow. The end result was just what I was looking for. From this angle, I could take the picture in the mirror and it looked like I was naked. I perfected my pose then snapped the picture on my phone. Todd would love it.

Before I sent it, I looked at it. I was surprised at how well I had done. The picture looked like it could go on one of those skanky calendars some of the guys liked. For a moment I considered not sending it, but in the end I did. It was only Todd, and he wasn't seeing any more skin than if I was in my suit; it just looked like it was more. And, okay, my pose was seductive, but since he was my boyfriend I wanted him to think I was sexy so he wouldn't look for another girlfriend. His reaction was exactly what I'd anticipated.

***OMG im saving that one***

***thats 4 ur eyes only***

***i know***

***promise!***

***yeah ttyl***

When I looked in the mirror, a big smile filled my face. Feeling like the luckiest girl in the world, I tried on the last swimsuit.

My happiness was short-lived. As soon as Danielle slammed into the pool locker room after school the next day, I knew there was trouble. She had a crazed look in her eyes like I'd never seen.

"What were you thinking?" she said through gritted teeth.

"What?" I'd had a good day and couldn't imagine what her problem was. Usually we got along well.

She extended a cell phone I didn't recognize toward me. "This is going all over school."

Feeling totally clueless, I took the phone from her. "Whose phone is this?"

"Connor Miller's."

I glanced at the phone. "Why do you have Connor Miller's cell phone?"

"Look at his wallpaper photo."

Horror shot through me as soon as the screen lit up. *I* was his wallpaper photo. But, not just any picture; it was the picture I'd sent Todd where it looked like I was naked. I gasped and fought the instant tears. "How did he get this picture?"

"Forget how he got it. What were you thinking texting a picture like this?" Danielle shouted.

"But, I sent it to Todd." Humiliation rolled over me in a sickening wave.

"Well, this isn't Todd's phone, is it Hannah? Do you have any idea how many phones and computers this picture is probably on right now?"

"What do you mean?"

"If Connor has this picture you can bet a lot of other guys do, too."

My breath caught in my throat like big hands were choking me. I could barely talk. "But why would Todd do that to me? Why would he send it to anyone else? He promised me."

"Ask him, not me." Danielle paced between the lockers and the wooden bench.

I fumbled through my coat pocket until I found my cell phone then dialed Todd's number.

He answered right away. "Hey, beauti-"

I cut him off. "Why did you do it?"

"Do what?"

Moments before I'd wanted to cry, but now I was so angry I wanted to climb through the phone and grab him. "You sent my picture to Connor!"

"I didn't send it or show it to anyone. I promised you that."

"Well, I'm looking at it on his phone right now." My voice was getting higher which meant at any second I would probably start crying. "How did it get there?"

"I don't know," he replied.

"You're a liar!"

"Hannah, I swear. I didn't send it to anyone." He sounded as upset and panicked as I felt, which didn't make sense.

Just then a few girls from my swim team came into the locker room. They were laughing and talking excitedly. I heard things like "sexting", "reputation ruined", "hottie" and other horrible words that by now I was sure were connected with me. When they saw Danielle and me they stopped abruptly and stared. They were my friends, my teammates, but they were acting like strangers. Courtney, who was the co-captain of the swim team with Danielle, came over to me.

"Sexting, Hannah? I didn't know you were into that."

I stared at her in disbelief. "I'm not."

Danielle stopped pacing and whirled around. "Leave her alone, Courtney."

Courtney wasn't intimidated and kept talking. "Hey, Hannah should know what's going on." She held her hands in front of her and studied her fingernails. "When a bunch of the guys were hanging out last night, Quinn picked up Todd's phone and found your picture. He forwarded it to a bunch of his friends who forwarded it to a bunch of their friends. And, well, I'm sure you get the idea. Unfortunately for you, you are legend. Doesn't look good for the swim team, though."

I wanted to melt into the floor and disappear forever. I couldn't even form words. Danielle came over and put her arm around me but that did nothing for the sick feeling inside me. She pulled the phone out of my hand, told Todd I had to go and guided me toward one of the bathroom stalls.

"What am I going to do? How can I ever face anyone at school again?"

Danielle stepped into the stall and ripped a handful of toilet paper off the roll then thrust it toward me so I could wipe the tears off my face. I sounded like a blubbering mess, so I'm sure I looked bad, too.

I leaned against the side of the stall and covered my face with my hands. "I wasn't naked. It just looks like it. And nobody except Todd was supposed to see it."

"The damage is done, Hannah. A picture like that shouldn't have been sent to anyone. Now it's in cyberspace forever."

Anger and embarrassment collided in my brain. If Todd was telling the truth, then Quinn was to blame for this horrible mess. How did I fix something like this? Retaliate? Find Quinn and scream at him? Wait and hope everyone would soon forget? More ideas rattled around in my head, but nothing made sense. There was no way I could undo the damage.

My teammates were back to talking among themselves. The locker room door opened again. The conversations stopped immediately.

"Is Hannah Lewis in here?"

I recognized Coach Parker's voice, and she didn't sound happy. Could I get kicked off the swim team for this? As her footsteps approached, I knew I'd have my answer soon enough.

# ISSC
# Internet Safety Savviness Challenge

Facts and opinions are two different things. A statement that is a fact can be supported with concrete evidence. A statement that is an opinion means that the answer is based on what someone thinks or believes, and there can be multiple answers.

<u>Directions:</u> Put an "F" on the line if the statement is a fact. Put an "O" on the line if it's an opinion. If it's an opinion, if you're reading this with someone else, take this opportunity to discuss your opinion.

1. ___ It is okay to take a friend's phone and look through it without his or her permission.

2. ___ If you send a picture, even one, and it is considered child pornography, it could result in legal action.

3. ___ If you send a picture to someone you really trust, no one else will ever see it.

4. ___ Only troublemakers would get themselves into this type of situation.

5. ___ Your reputation can't be affected by one incident.

6. ___ A picture that is sent over a cell phone could be transferred to the Internet.

7. ___ In the long run, your actions and decisions only affect you.

8. ___ Sometimes kids will forward a questionable picture sent by a girlfriend or boyfriend they've broken up with if they are angry and want to retaliate.

9. ___ Texting a "mature" picture of yourself to a boyfriend or girlfriend is proof of how much you love them.

10.___ If someone forwards an inappropriate picture to your phone you are responsible for the consequences if someone finds it on your phone, even if you didn't ask for it.

# What can we learn?

Hannah's actions put herself, and possibly other people, into difficult situations. The act of taking a picture and sending it to a friend is not criminal or even wrong, but once those photos become "more mature" or "inappropriate", there could be very serious, long-lasting consequences.

***"I wasn't naked. It just looks like it."***

In Hannah's opinion, her picture was innocent. Hannah felt this was okay, but what she did not think about was how other people would perceive the picture. If it looks like you are naked in a picture, people are going to think you are. Hannah trusted Todd and Todd did not do anything to break that trust; however Todd's friends did. When Quinn picked up Todd's phone and looked through it, this is when the real trouble started. It's up to each individual to decide whether they consider it an invasion of their privacy if a friend goes through their phone. A good friend would ask first.

The embarrassment that Hannah will face is going to be very difficult to deal with, but what if that photo was a little more "mature"? If it fell under the rules of pornography, a whole new problem would have begun.

First, Hannah produced a photo that could be considered child pornography even though it is a picture of herself. Because she's under 18, she's still considered a child. That is a serious criminal offense with harsh penalties. Then the problem gets bigger after she sends it to Todd because now he is in possession of an image of child pornography. In some states with certain laws he could be known as a sex offender even though he really didn't do anything wrong. Finally, Quinn or anyone else who forwarded the picture could be classified as someone who distributes child pornography and may have to register as a sex offender, as well.

When these types of incidents occur, nothing good ever comes out of them. First, young people shouldn't see or possess photos like this, but when they share them with people or post the photos on the web, they can wind up in serious trouble. Students who are looking at Hannah's photo are part of the problem, and students who bring the photo into school are not only part of the problem but could also face serious school consequences.

If Hannah signed a contract with her swim team saying that while on the team she will not conduct herself in any manner that would embarrass the team, now she could lose her position. How is Hannah going to feel when her parents see the picture? (And they will see it.) One of Hannah's true friends will tell their parents that Hannah may be in trouble. Those parents will call Hannah's parents, which is the right thing to do, but that will mean that Hannah's parents will see their daughter in a way that no parent wants to.

Before anyone sends any image they should ask themselves, "Do I care if everyone sees this picture?" If the answer is "yes, I do care", DO NOT SEND IT. Hannah's situation proved that, although we can trust some people, there are people you can not. Situations like Hannah's can get out of control very quickly.

Hannah will have to deal with this situation for a long time. For the rest of her high school years students will probably talk. New kids to the high school may be told, freshman girls new to the swim club will hear about the "Hannah Rule" (no sexting), members of the teams for schools who compete against her in swimming may make comments during meets, and teachers may also look at her differently. One day Hannah could be in a college classroom when someone behind her quietly asks her if she is still swimming and are bathing suits required? This one lapse in judgment will require a long healing process that could have been avoided.

Another element to consider in this scenario is how this affects Hannah's sister, Danielle. Danielle already faced a problem in the locker room when the other girls gave Hannah a hard time. Danielle wanted to protect Hannah but was ignored. Danielle will hear about this too. Teachers will ask, "Is your sister ok?" Fellow classmates will ask, "What was she thinking?" And some boys might ask, "So, is posing for pictures a family tradition?" This incident will have long-lasting effects on Hannah *and* her family.

The problem with this type of situation is that it is almost impossible to completely get rid of the picture. If the picture is not child pornography but is embarrassing or something someone is ashamed ever happened, there is not a lot legally anyone can do. Parents, if your child ever does do something like this, one approach would be to contact the person who first started to circulate the picture, if that can be determined. This obviously will not get every picture deleted but would show that parents are willing to communicate and work together to help protect kids. It will also send a message in your community to other kids that this type of behavior will not be tolerated in the future.

The school system should also be notified and brought into this situation. They may not take disciplinary action, but they should be advised due to the fact that this information will spread through the student population very fast. School staff could hear more information from kids about who is circulating the image, if not known, and they might also hear if any web sites are now posting the picture. If school administrators, teachers and coaches are unaware of the situation they might not put the pieces together that someone may need help.

The school may also deal with the results of this incident. If Todd confronts Quinn in school, the history of the event will help the school deal with it and keep everyone safe. School systems also may have education programs or speakers who can come in and address the issue with the student body to help prevent future episodes.

If the image is pornography, or worse child pornography, contacting your local law enforcement is important. With child pornography being distributed, it can be a very scary and difficult situation. If a child keeps a photo on his or her phone, knowing it is child pornography, that is possession. If a child receives these images from a friend, and did not ask to receive the image, the child who received it did nothing wrong. But, if he does not do anything about the situation and keeps the picture, that could be an illegal act.

This is why kids need to understand why it is so important to talk to their parents. The best approach a child could take is to tell his parents immediately that he received an image. The parent should then contact law enforcement as soon as possible, if not immediately, and explain the situation.

No kids want to tell on other kids, especially if they are friends, but in this case, telling an adult or law enforcement official is protecting yourself.

# ISSC Answer Key

1. Opinion
2. Fact
3. Opinion
4. Opinion
5. Opinion
6. Fact
7. Opinion
8. Fact
9. Opinion
10. Fact

# CHAPTER 11

# Michael's Mistake

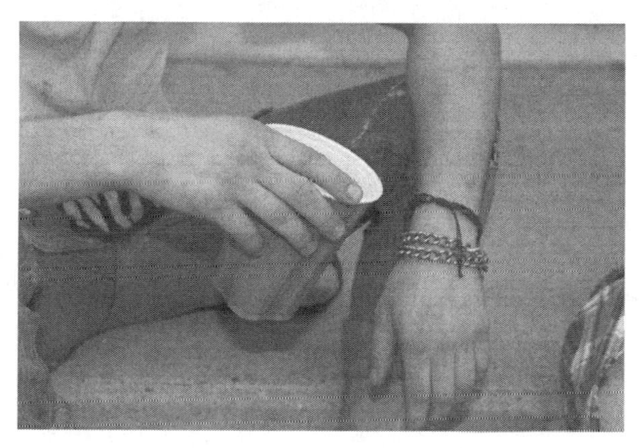

**Profile:**
17 years old
Scholar Athlete: soccer and baseball
Mother: Nurse
Father: Engineer
Siblings: 20 year old sister, 21 year old brother
(both in college)
Town demographics: 58,000 residents

My hands were still shaking when I tossed the letter from TSU aside and signed on to the Internet. I was pumped and couldn't wait to share the good news. One click and I was on *FriendPost* to share my news on the message boards.

> ***Michael Berlussi: It's official... playin' ball for the Panthers next year.***

With my older brother and sister in college, in order to go to TSU, I had to get a good scholarship. Although I'm good in academics, I knew a baseball scholarship was my best hope. I picked up the letter again and stared at the words I'd been waiting to read for months. This called for serious celebrating with my friends, and the timing couldn't be any better. Tomorrow my parents were flying to Georgia for a wedding. I'd have the house to myself for the entire weekend.

I switched over to the browser and pulled up the bookmark for TSU. When the photograph of the majestic brick building loaded, the excitement shot through me again. This was a dream come true. I spent a few minutes navigating the school's website, trying to imagine what my life there would be like. A flash in the lower part of the screen alerted me to incoming posts. I clicked over to *FriendPost* and saw several comments from friends.

***D.J. Werkmen: awesome dude! full ride?***

***Meghan Burkewall: sweet. can't wait to visit u at TSU***

***Dustin Nord: road trip man!***

***Lisa Tambiene: pa-a-a-r-r-t-t-t-ty!!!!!***

I added to the comment thread.

***Michael Berlussi: Yeah. Full ride for baseball***

***Meghan, come on up. It's a great campus.***

I heard the garage door opening and knew Mom was home. I met her at the kitchen island. She set down bags full of groceries.

"You're looking happy," she said. "What's up?"

"Check it out." I handed her the letter. Her eyes grew wide and the biggest smile I'd ever seen spread across her face. "Sweet, huh?" I said while she read.

She came around the island and hugged me. "Michael, this is fantastic." Her voice sounded raspy like she was going to cry. Great! Just what I needed was her getting sappy.

I let go of her and held the letter out at arms' length. "I know. Four years of college for free." I pulled the letter back toward me and kissed it. "Ah, can't beat that deal."

Mom laughed. "I'm so proud of you. I bet you're excited."

"You have no idea."

"Yes, I do." She tipped her head and winked. "I know it's hard to believe, but I was trying for college scholarships once, too. And, despite what you kids think, it wasn't that long ago." She laughed. "After Dad and I get back from the wedding this weekend we'll take you out to celebrate all the money you're saving us." She started putting groceries away, so I helped. "I got you some easy meals to have this weekend." She held up a frozen pizza before putting it in the freezer. "Supreme. Your favorite."

"Nice. Thanks, Mom."

When we finished I returned to my computer. There were a lot more congratulatory comments on my *FriendPost* page. If I were walking on clouds I couldn't have been any higher. I was invincible. I clicked off that site and moved over to my *MesssageMe* profile. Dustin and I had already planned a poker party for Saturday night at my house, but I saw that he'd turned it into a different kind of party thanks to my news. I had to laugh. Dustin had been bugging me that we should invite more people since we had the whole house to ourselves, but I'd held off. I could only afford so much food and drink. But, he had already taken matters into his own hands and posted an invitation on *MessageMe*.

***Dustin: Party at Mike's Saturday night. 'rents gone for weekend. Gotta celebrate scholarship. Who's in?***

***Lisa: me what time?***

***Bob: be there***

***Rick: in***

***Rochelle: count me in***

***Chris: on it***

Those were only the first comments. The responses seemed endless. I stopped counting after twenty. It felt good to know that many people wanted to come to a party I was having. Maybe I'd beat Craig Messier's guest list that topped fifty a few weeks ago. Everybody was still talking about that party. With our family room and the back yard, as long as the weather was good, we'd have no problem with that many people. I figured I better make sure people knew to bring food and drinks with them because I didn't have the money for a party this big.

***Mike: Will need help with "refreshments". BYOB or chip in. Time to celebrate!***

Almost immediately Dustin posted another message.

***Dustin: i'll have my brother get a keg***

I wasn't so sure about the keg. Having a few bottles of beer around was one thing, but a whole keg at our house – well, I wasn't sure how comfortable I was with that. Yes, there could be a lot of people at the party, but it seemed like a keg would invite over-drinking for some people. I was responding to suggest we hold off on the keg when my mom called for me.

"Mike, can you help me with the suitcase?"

"Be right there." I stared at Dustin's message again then changed my mind. I'd probably never have another opportunity to have a party like this. Having a keg would kick it up a notch. I clicked the mouse and sent off the message ***you rock!*** then minimized the screen. This party was going to be unforgettable.

The next day my parents were on their way to the airport before I left for school. They gave me money in case I needed anything before they returned late Sunday night. I could use the funds. Once I got to school, everybody was coming up to me to say they'd see me Saturday night. All of a sudden fifty sounded like a small gathering compared to what was cooking at my house.

Saturday, shortly after noon, Dustin's dad called. I'd always thought he was a pretty cool guy, but by the time I hung up the phone there was no doubt.

"Mike," he said when I picked up, "I know a few of you boys are getting together tonight to, uh-" he cleared his throat and chuckled before continuing, "-to play a little *poker."* The way he emphasized the word "poker" I could tell he was aware of the plans. "I have a suggestion. Take everyone's keys when they come in the door and only give them back when they're leaving. That way you can make sure they're not *too tired* to drive, if you know what I mean."

I knew exactly what he meant and wished my dad was as relaxed as Dustin's. "Thanks," I said. "That's a great suggestion." When I hung up, I nodded to myself. That sounded like a good plan, although I didn't know what I'd do if the neighbors saw a bunch of cars in our driveway all night.

I got back on *FriendPost* and sent off a message.

***Michael: See everyone after 8. No one parks in our driveway tonight. Car pool so not many cars r on r street***

Dustin rolled in with the keg before 7 p.m. It was going to be a warm early spring night, so we set up tables outdoors to handle the overflow of people. It didn't take us long to get the place ready. I was jittery, partly from excitement, but partly because I was worried that my parents might find out. Our neighbors were far enough away that there'd have to be a real lot of noise for them to hear us, but I couldn't control what they could see.

By 9 p.m. the party was in full swing, and I was feeling good. Our house and yard were packed with kids from school and even kids I didn't know from other towns. It was obvious a few of them had already been drinking before they arrived. As Dustin's dad had suggested, I took all of the car keys that I could as people showed up, and I put them in a big box in the kitchen. I thought that was pretty smart because if they were too drunk to figure out which keys were theirs then they shouldn't be driving.

Our house was pulsing with the loud music, the kids trying to shout over each other to be heard and laughing. There were some kids at the party who weren't drinking, I'm sure, but it was hard to tell since everyone had cans of beer or blue plastic cups in their hands with something in them. I decided to play it safe and just carry around a cup with cream soda in it to make it look like I was drinking. The party had seemed like a great idea when we planned it, but I had to admit now that it was actually happening, I wasn't as comfortable. There were a few kids who were getting so drunk that they had to run outside to throw up. I hoped no one got sick in the house or had any big spills.

Jason Walker, the catcher on my baseball team, stumbled up to me, his arm hung over a pretty girl I didn't recognize.

"Great party, man." He nodded toward the girl with him. "This is Claire Sheffield from Warren High. We hooked up here, man. She's awesome. Thanks."

Claire smiled at me. "I hope you don't mind that a few of my friends and I crashed your party."

"No way! You know the saying, *The more, the merrier.*"

She glanced around. "There're definitely more here." Jason started kissing her ear. She didn't seem to notice him doing it.

"I heard about your baseball scholarship," she said. "Congrats. My brother is a freshman at TSU and loves it."

"Thanks. Really, that's cool. Maybe we'll end up having classes together."

"Yeah, maybe."

We stood there awkwardly before Jason finally quit nuzzling her and looked in her cup. "You're empty. Better go fill 'er up."

There was a slight slur to his words, and the way his voice went higher on the word "up" made us laugh.

"Congrats, again. Great party," Claire said as he led her away.

"Thanks."

I watched them get swallowed by the crowd. Not everyone was drinking a lot, but there were a few, like Jason, who definitely would not be getting their keys soon.

Throughout the night a lot of people came and went. There were kids I didn't know and never talked to the entire time they were there. It didn't matter. Everyone was having a great time.

About an hour later I was sitting on the couch with Randy and Dustin debating the standings for the teams in our division when

Claire came back with a friend. The girl had long, wavy hair and the most gorgeous big brown eyes I've ever seen on any girl.

"Hey, Mike," she said to me, "this is Artemis. She's a foreign exchange student from Greece. She wanted to meet you."

I slid over on the couch and patted the cushion. "Well, come on over and meet me, Artemis." Since I had broken up with my girlfriend a month before, I was a free agent.

She giggled and shimmied in between Randy and me. "Great party, Mike," she yelled over the noise. She had a cute accent that made her seem even more exotic.

"Thanks. I'm glad you and Claire came."

"Jason said you are captain of the baseball team."

"Co-captain," I corrected. "Kirk Somers is the other one."

She nudged me with her elbow. "Still a captain. I will have to come to a game."

"How about a picture?" Claire asked, holding up her cell phone. "Artemis can add it to her scrapbook about her year in the U.S."

I glanced at Artemis. We shrugged at the same time, which made us laugh, and then we held up our blue cups. Claire's flash went off a couple of times. "Perfect!" she declared, and we put our cups down.

"You two have fun," Claire said. "I'm going to find Jason."

So there I was, stuck with a beautiful girl and not minding a bit. We hit it off right away. It didn't take long for me to decide that Greek girls were hot.

It was almost 1 a.m. when Artemis, Claire and the rest of their friends decided to leave. The party had wound down quite a bit, and there were only a few people still hanging around.

"I'll walk you to your car," I offered. I wanted to get Artemis's phone number before she left. Jason walked out with Claire, if you could call it walking because I wasn't sure who was holding up whom.

"Dude, how are you getting home?" I asked him.

"Claire's droppin' me off." His speech was slurred, so I was relieved he wasn't thinking he was driving anywhere.

"Are you sure you're okay to drive?" I asked Claire.

"Nikki's going to drive." A blonde girl held up the keys and jingled them for me to see. "She's only had a couple drinks tonight. She said she's okay to drive."

Nikki looked like she was walking okay. Claire's little Ford Escort was parked around the corner. It was dark there, so I was glad I had walked with them. While the others got in the car, I pulled Artemis aside.

"It was fun getting to know you." I took both of her hands in mine and bent my knees so I was at eye level with her. "I'd like to see you again sometime. Would you like that?"

There was enough light from the moon for me to see her wide smile. "Yes, I would like that."

I pulled my cell phone from my jeans pocket and took her number. "I'll call tomorrow, okay?"

She tipped her head like she was suddenly shy. "I would like that."

I gave her a quick kiss then helped her into the seat behind the driver. When I closed the door, she smiled up at me. Yep, this was definitely an awesome party. Nikki started the engine and pulled away. I couldn't wait for tomorrow.

Fortunately our house and yard were still in good shape despite the party, but a few of my friends helped me finish the clean up. After they left, I checked on-line to see if Artemis had a *FriendPost* page. Her last name had a kazillion letters in it, like many Greek names. Finally I found her, clicked on her profile and sent a "friend" request. I looked at the "newsfeed" and there were already pictures from the party up because people had posted from their cell phones during the party. I had been linked in two pictures, the one with Artemis and one with a couple of guys from the baseball team. The blue cup I held seemed to jump off the page. What a party!

I had planned to sleep in late Sunday, but a little after 8 a.m. my cell phone rang.

"Mike!" It was Dustin and he sounded panicked. "Dude, you gotta turn on Channel 7 News quick."

"What?" I rubbed the heel of my hand into my eyes to encourage them to open.

"There was an accident. Turn on the news."

I rolled over in my bed and grabbed the remote for my small television. My eyes didn't want to adjust to waking up so early and so fast. I clicked on Channel 7 and the image that filled the screen made me sit up straight. There was a picture of Claire Sheffield's Ford Escort, but the whole driver's side of the car was smashed against a tree.

"What the -" I was too shocked to finish my sentence. In my mind all I could picture was closing that back door after Artemis got in and her smiling up at me. "Is everyone all right?"

Dustin didn't have time to answer before the reporter stepped in front of the camera, her grim expression telling the story.

"The occupants of the car have not been identified, Stan, but we have been told that there were five teenagers in this vehicle. Although authorities haven't released a statement, friends of the teens at the scene have told us that the driver and the passenger behind the driver sustained life-threatening injuries."

My chest tightened and I thought I'd choke on the breath I'd taken. My arms and legs started shaking. "I gotta go," I told Dustin and pushed "end" on my phone. The news camera focused one more time on the mangled car, and all I could do was stare at the wreckage, wondering if Nikki had been drunk after all.

For the rest of the morning my phone never stopped ringing, and the *FriendPost* site was filled with comments and conversation about the party and accident. No one actually came out and said the accident happened because of the party, but a lot of people were questioning whether Nikki was drunk.

By mid-afternoon Dustin and our buddies, Tommy, Rico and Jon were gathered at my house. There was no way my parents weren't going to find out about the party now. I paced like a caged animal as the guys tried to get more information about Jason and the other two in the car. The text messages flew, and I monitored the *FriendPost* site. There were rumors flying over the Internet, including one that someone had died. I was sure I'd go crazy before the end of the day.

A little after 4 p.m. Dustin received a text from Jason. We crowded around him and his cell phone.

***home from hosp broken arm/ whiplash hangover worse lol***

"Ask him about the others," I said. Dustin typed in the message and received an immediate response which he read out loud.

***Nikki & Artemis in hosp. Not good.***

The guys all glanced up at me then the phone beeped with another message.

***Claire n Nikkis bf home. Cuts.bruises only***

I dropped into the recliner next to the couch and stared at the clock. My parents would be home in four hours. If Friday had been the best day of my life, today definitely came in as the worst.

Dustin, Rico, Tommy and Jon left by 6 p.m. By then we'd had more news on Artemis and Nikki and their conditions had been upgraded a little. I'd quit answering my cell phone, and I couldn't go near the computer.

A little after 8 p.m. a car rolled up in front of our house. My parents had called a little more than a half hour before to say they were leaving the airport. My heart pounded and sweat beaded on my palms. I'd never let them down the way I had this weekend, and they didn't even know it yet.

When the doorbell rang I nearly jumped through the ceiling. Why didn't they come in the garage? I went to the front door and peeked through the side window. Two uniformed police officers stood on the front steps. Simultaneously, my parents' car swung into the driveway and came to an abrupt halt in front of the garage. It had barely stopped when my parents leaped out and rushed toward the policemen. I swung open the door to face them all.

"Is something wrong?" my mother asked since she was the first one to them.

"Mr. and Mrs. Berlussi?" one officer asked. His nametag said Wilson. The other officer's name was Mielzinski.

"Yes," they answered at once.

Officer Wilson then turned toward me. "Michael?"

I nodded.

Officer Mielzinski held up a large manila envelope. I could only imagine what was in there. "We need to talk."

The look of alarm on my parents' face killed me. I backed into the foyer and swung open the door so they could all join me inside for my nightmare.

Dad invited the officers into the living room but they opted to stay in the foyer. The officers explained about the accident during the night while I stared at their feet.

"What does this have to do with us?" Mom took a step closer to me so our arms brushed against one another. "Or Michael?"

Officer Mielzinski removed a handful of photos from the envelope and handed them to my mom. "These photos were posted on *FriendPost* profile pages of several kids last night."

Dad stepped next to Mom and they studied the photos as my mother flipped through them. There were pictures of some kids playing a drinking game, a picture of a whole row of guys with beer bottles tipped up and almost empty, several other pictures of kids obviously drinking and then the picture of Artemis and me holding up our blue cups.

Officer Wilson cleared his throat then asked, "If you look at the background in these pictures, would you say they were taken here at your house?"

Mom and Dad stared from the pictures, to me, back to the pictures and then glanced at the rooms behind them as if they needed confirmation that they weren't imagining things. Finally, Dad's steely stare landed on me. His mouth was a hard, straight line.

"This looks like our house, but we were in Georgia for a wedding this weekend."

"And where were you?" Officer Mielzinski asked.

My face was so hot it had to be redder than our school's scarlet team uniforms. "I was here," I choked out.

Everyone stared at me until our thoughts were interrupted by an SUV pulling into our driveway. We all turned toward the bay window and stared toward the driveway. When Coach Canata and the high school principal, Mr. Potterstein, emerged, I knew immediately that the kids in the accident weren't the only ones whose futures would be affected because of this party.

# ISSC
# Internet Safety Savviness Challenge

Answer true or false to the following comments. Check your answers at the end.

1. On a social networking site, if I have my account set to private, only my friends can see pictures of me.
2. Everyone who uses social networking sites must use a real picture of themselves.
3. Social networking sites can be "hacked" into just like other web sites.
4. If I post a picture on my social network site, that is something that can not hurt me in the future.
5. People can download and take possession of my pictures if they are on social networking sites.
6. Once I set my privacy settings on social networking sites, they never change.
7. My actions in regard to Internet use can potentially affect my family and friends.

# What Can We Learn?

Social networking sites have become a huge part of society. Millions of people are on them - parents, grandparents, students, news channels, restaurants, local government, celebrities, and the list goes on. As a result, the size of the world community with which we connect via the Internet is mind-boggling.

Of course, not everything about social networking is bad and there are some benefits to them. The problem is sometimes we forget about safety when we're sitting in our homes or using our cell phones, and as a result, we put ourselves in bad situations. We *must* be aware of what we are doing and posting at all times because things happen fast and get out of control quickly on these sites.

Michael and his friends made many poor choices in this situation. They put themselves and everyone around them in danger. In Michael's situation, social networking sites had both negative and positive results.

The negatives:

Having police officers come to Michael's door can only mean one thing; he is in BIG trouble. Because the kids were at Michael's house, and the social network site photos prove that, Michael can face some very serious charges. Each state has different laws, but some of the possible criminal charges could be: hosting an underage party, providing alcohol to a minor and risk of injury to a minor. In some states Michael could be charged separately for each teen affected, meaning if it was proven that 15 kids consumed alcohol while on his property he could be charged with 15 counts of providing alcohol to a minor. There is no doubt of what was going on at the party and who was there because the pictures tell the story. The evidence of crimes will be found.

The pictures that were posted not only led to the police involvement but those photographs are now a permanent part of these kids' past. Once posted on the Internet, these photos will never disappear. Future bosses will be able to see what type of character a person is by pictures posted on the web. Colleges and scholarship boards will be able to, or may request as part of the application process, to see your social networking sites.

If you are applying for a scholarship, you want every advantage possible. If there are several candidates applying for the same scholarship and you're the only one with photos on the Internet of you drinking alcohol, where do you think you will fall on the list of candidates? Across the country schools and employers are taking advantage of the background information they can gather from the Internet to help them choose the right people for their school or company.

Teens don't often consider the consequences their actions may have on their parents' lives. Let's look at Michael's parents. When they left Michael alone, they trusted their son, something every parent and child wants. As soon as they saw the photos of what happened in their home, trust was compromised. Michael's parents will be hearing a lot about this situation, not just from the police and attorneys, but from their neighbors, friends and co-workers. They will have to live through this ordeal for a long time, and all they did was trust their son.

The possibility of even more devastating effects for Michael's parents is illustrated by a story told by a woman who called into a radio talk show. She informed the host that she had recently lost her homeowner's insurance because her agent saw her child's social networking site. In the agent's opinion, the behavior that was demonstrated on the site was risky and cause to drop the woman's policy. The consequences of kids' actions don't stop with them. Parents have to go through all of these incidents, and deal with them, alongside their children

The positive:

Although some may disagree, it's actually a positive that, because of access to the social networking site, law enforcement and the school system were able to find out who was responsible for the crash that damaged so many lives. If you look at it in this light, it was a positive that people posted pictures of underage drinking and unnecessary risks being taken. It was a positive that someone did the right thing and turned over the information they saw on the site to law enforcement and the school. Another positive regarding the posting of photos and comments is that, perhaps because of what people see or read, they may think before they put someone in danger with reckless decisions.

The focus of this book is Internet and technology safety, but we have to address the other big safety concern that arises in this chapter: underage drinking. When teens host a party at their house and underage drinking occurs there are several ways this can go very badly.

First, as we said earlier, because it is against the law to provide alcohol to a minor, the hosting teen is now violating laws. This, too, can become part of your history and permanent records that you would potentially have to disclose to future employers or schools. Having a police record isn't a positive way to enter adulthood.

Another concern is in regard to what happens if kids get sick from alcohol poisoning. Will the other teens do the right thing and get help or will they just let their friends "sleep it off"? If a friend lets someone "sleep it off", the ramifications of those actions can be costly. Kids across this country have died from alcohol poisoning and overdose when medical attention was not provided when it should have been. Both girls and boys have become victims of assaults, physical and sexual, when they have lost the ability to make their own decisions or stand up and protect themselves from others. No one who drinks to excess is immune to these dangers.

Finally, regardless of whether or not the parents are aware of a party in their home, they can be held civilly liable for things that occur on their property. Michael's parents had no knowledge of what was happening, however, they will probably be held accountable.

Underage drinking is not a rite of passage; it is dangerous. As we learned from Michael, the consequences can be very severe and life-changing. Life-changing consequences also occur when we put our lives on social networking sites. Images you post are a permanent part of who you are and what people can learn about you. What might seem fun, silly or "cool" now, might not in a few years. Instead of being cool, your posts may jeopardize your chances at a career or opportunity. Is it worth it?

# ISSC Answer Key

**1.** FALSE- Other friends can post your picture. On some sites you can be "tagged" in a photo. "Tagging" identifies you in a picture and allows people, who may or may not be a friend, to see it.

**2.** FALSE - No social networking site verifies pictures.

**3.** TRUE - Social networking sites, like many other sites, are vulnerable to hackers.

**4.** FALSE - It has been proven time and time again that pictures and information on the Internet can hurt your future.

**5.** TRUE - Almost any image on the Internet can be downloaded when you view it

**6.** FALSE - Social networking sites can change security/privacy settings at any time. If you are not on top of it and make necessary changes, your information can be unprotected.

**7.** FALSE - If you post pictures or comments that imply criminal action and you're a minor, your parents could be held responsible if your actions cause injury to someone else.

# CHAPTER 12

## Olivia's Ordeal

**Profile:**
16 years old
Assistant Editor of the Yearbook, Cheerleader, Soccer player
Mother: Public Relations Director
Father: Middle School Math teacher
Siblings: sisters 14 and 11
Town demographics: 2,300
School: K-12 Central School

"Okay, gang, we're on deadline." Mr. Rice, our yearbook staff advisor, looked serious. "One month from now we have to submit the final pages to the printing company. To accommodate your activity schedules, for the next four weeks we'll have two or three meetings after school and one before school every week. You'll need to be at every meeting, if possible. The first one will be tomorrow at 6:45 a.m."

Most of the kids groaned. "It's once a week," he said. "You'll survive."

He turned toward Heather Wheaton, the senior editor, and me, the assistant editor. As the ones "in charge", we sat at two desks facing the rest of the yearbook staff. Mr. Rice sat on a student desk adjacent to us and continued.

"I'm leaving it up to the two of you to organize the staff into mini task forces that will tackle each aspect of the jobs that need to be completed."

Heather and I glanced at each other and nodded. "We're on it," Heather assured him.

One of the things we like about Mr. Rice as our advisor is that he lets us handle most of the work. Since our school is so small, there are only seventeen kids on the yearbook staff. Still, it's fun to have others look up to us as the ones in charge.

Mr. Rice slapped his hands on his knees and looked around the room at the committee. "Well, get to work."

First, Heather and I divided the staff into the task forces. Heather took the layout crew, and I took the photographers. There were three: Fiona and Aaron, seniors who had been on the yearbook committee since their freshman year, and Will Barnard, a sophomore who transferred to our high school at the beginning of this school year. Will joined the yearbook staff because he was in Mr. Rice's technology class, and Mr. Rice encouraged him to join as a way to meet other kids.

High school can be tough, and if the "in crowd" decides you're a target, you aren't going to have it easy. For whatever reason, whether it was his accent because he was from another part of the country, or because he dressed a little differently, from the first day of school Will had been a target and was teased a lot. Some of the upper class boys called him "Barnyard", and when they passed him in the hall they made farm animal sounds. It encouraged them even more when other kids in the halls laughed at their behavior or joined in.

My boyfriend, well, *ex-boyfriend*, Paul, was one of them. Once, at a basketball game, Will was leaving the gym and Paul stuck his foot out and tripped him. Will fell into our cheerleading squad.

After he got up, he turned bright red, hurried out of the gym and disappeared down the hall. I never saw him come back into the game, so I assumed he left, humiliated.

Except for the fact that Will was real quiet, to me he seemed okay. At yearbook meetings Aaron and Fiona paired up for photo assignments and left Will to work by himself. He didn't seem to mind, but sometimes I thought it was sad that seven months after coming to the school he still didn't fit in anywhere.

I started to get up from the desk when Fiona approached and held up a camera. "Aaron and I have the club pictures. We'll download them onto the computer, sort through them and let you know when we have something to show you."

"Okay." Even though I was the assistant editor, they were upperclassmen, so I didn't argue. Fiona pivoted and followed Aaron to a computer on the other side of the room.

That left Will and me. I pulled a chair up next to the computer he was booting up.

"Show me what you've got."

While he stared at the screen, I looked at him from the side. Red streaks crept up his neck and across his cheeks. Had I embarrassed him in some way? Then I remembered a conversation with my friend, Cindy, a couple of months before and realized what was happening. She told me she thought Will had a crush on me because she always saw him staring at me from across the lunchroom. At the time I laughed it off and told her she was crazy. Now I wondered if she was right. This kid's blush could start a fire.

My attention was drawn back to the computer screen filled with thumbnails of hundreds of candid shots Will had taken during after school activities, in the lunchroom and classrooms. When I enlarged them, I was stunned by his skill. Not only had he captured action, but he'd also nailed the emotions behind each shot.

"These are incredible." There was nothing fake about my compliment. This kid had serious talent. "Where did you learn to take photographs like this?"

He shrugged and ducked his head, obviously not used to compliments. "I used to go to camp every summer, and they offered photography classes. And when I was 12 my dad signed me up for a class at the community college. I've been taking classes ever since."

"You pay attention in your classes. Man, you rock!"

A smile lit his face. "Thanks."

We continued clicking through what he'd taken, moving the ones we wanted to use in the yearbook into a separate file. After an hour of work we'd also decided on which pages we'd use them.

I sat back in the chair and hit Will's upper arm with the back of my hand like we were buddies. "You know what we need? Candids from *our* meetings."

I snatched the camera he'd laid next to the computer and turned it on. "Let's start with us, the assistant editor and the photographer working hard at a meeting." I flipped the camera around so the lens was pointing at us and leaned toward him. "Say cheese!" I clicked several pictures. At first he was stiff, but after seven or eight shots, Will surprised me by throwing his arm across my shoulder and smiling. When we previewed the photos, we laughed out loud at some of them.

I turned and gave him a high five. "This was fun. Really, Will, you rock!"

Will's assignment for the rest of the meeting was to take candid shots of the staff while I worked with Aaron and Fiona. Much later, when Will was leaving, I made a point to say goodbye.

"Hey, Will." He turned in the doorway to face me. "Can you make sure you come to the before school meeting tomorrow to work on the new candids?"

He glanced around like he was sure I had to be talking to someone other than him. "Uh, sure."

"Great! See you then."

He lifted his hand in a half wave then spun toward the door as if he had to hurry out. Even his posture seemed straighter like he was pumped. It made me smile, too. I liked being in charge.

It wasn't until a couple of days later that I regretted my brilliant idea about the candids. I received a text message from Paul.

***thought you wanted to take a break from dating***

***yeah?***

***didnt know you were into farmers***

***what?***

***check out Barnyard's profile pic on FriendPost.***

My first instinct was to tell Paul how juvenile he sounded when he called Will Barnard that name, but I was curious about why he wanted me to check out the picture. I signed onto the site and searched for Will. As soon as his profile loaded I was surprised by what I saw and also knew what Paul's issue was. Will had used the candid of the two of us with his arm around me and made it his profile picture. I saw Paul was currently on-line, so instead of texting, I clicked on chat.

***Olivia Dawson: that pic was taken at our yearbook meeting So what?***

***Paul Heberfield: cozy barnyard is mooooving in on my girl lol***

***Olivia Dawson: I'm NOT ur girl. and ur being ridiculous. Now u know why I needed a break***

***Paul Heberfield: yeah so u could move on to that loser***

***Olivia Dawson: get a life***

Before Paul could respond I clicked out of the chat and returned to Will's profile. Really, the picture was harmless. In fact, it was surprising to see him smiling because he never did in school.

Word about the profile picture spread fast. The next night my friends, and even kids I barely knew, were hassling me on-line.

***Shawna Richards: everybodys checking out Will Barnards profile pic.***

***Penny Lane: new b/f?***

***Ellie Rigby: junior prom date for sure***

***Olivia Dawson: lol It's just a pic from yearbook mtg.***

***Lucy Diamond: he has his arm around u!!!!!! ewwwwww!***

***Olivia Dawson: We were goofing around at our meeting. Doesn't mean n e thing***

***Michelle Mibella: i cant believe paul feels threatened***

***Olivia Dawson: Paul's being a jerk. He just doesn't want to admit we're through.***

***Olivia Dawson: everybody's getting worked up over a profile picture and hassling me***

***Shawna Richards: tell him to take it down***

***Olivia Dawson: why? He likes it.***

***Madonna McCartney: whatevah***

***Olivia Dawson: yeah whatever lol***

I thought that was the end of the ordeal, but a couple of days later I realized it was just the beginning. Chris Porter, one of Will's few friends, stopped me on my way into calculus.

"Olivia, Will needs your help," he said.

I glanced around. Will was nowhere to be seen. "What do you mean? What kind of help?"

"Heberfield has kids ganging up on him everywhere he goes. Some are chasing him off school property and throwing things at him. They're telling him he has to change his profile picture."

I shifted my books to my other arm. "So why doesn't he?"

"He says he's tired of being bullied."

I thought Chris's accusation was a bit dramatic. "The kids are just teasing. Lots of kids get teased."

"Olivia, it's not just teasing. It's worse than it's ever been. He's started skipping classes because certain kids are in them, and he doesn't eat in the cafeteria anymore."

I realized he'd also skipped the last two yearbook meetings even though we're on deadline. I hadn't given it much thought because I figured we were down to mostly layout and there were no more photos to take. For a moment I felt bad that I'd hardly noticed his absence, but then something inside me snapped.

"Wait! None of this is my fault," I countered. "I can't control what Paul or the rest of the kids do."

"But Will's really down. I don't like the way he's talking. Have you seen what he's posting on his bulletin board at *FriendPost*?"

"No, I'm not his friend." That sounded cruel, so I added, "On *FriendPost*."

"You should *friend* him and see. It's dark stuff, and it's gotten worse. I'm worried." Chris reached out and gripped my arm. "He thinks you're really nice. Maybe you can convince him to change the profile picture so they'll leave him alone. Then maybe it won't be so bad for him."

Annoyed that I was dragged into this drama, I pulled my arm out of Chris's grasp. "Fine. I'll suggest it."

Even though I was concerned about sending Will the wrong message about my intentions, that night I sent a *friend* request to his *FriendPost* page. He was still using the photo from the yearbook meeting as his profile picture. He must have been on-line because he accepted the request immediately. I clicked onto his *bulletin board* to see what Chris was talking about.

There were conversations with kids I didn't recognize, so I guessed they were from his old high school. I skimmed his older posts and realized how isolated he felt. In conversations with some of these people he talked about school being a worse hell than what he would experience in death. He used lyrics from songs and poems about death and escaping torment to make his point. Even though he'd been posting things like that for a while, I could see that in recent weeks the lyrics were getting scarier. I also noticed that the day he posted the profile picture, and for the next day, his posts were unusually upbeat. Shortly after, his posts plummeted into darkness again. I guessed it coincided with the teasing at school.

I sent him a private message rather than post on his bulletin board.

***I'm getting hassled about your profile picture. Change it.***

Satisfied that I'd done what Chris asked, I clicked off Will's bulletin board and returned to the live feed. I was startled to see that several kids had joined a group called *Supporters of Will Barnard's Sex Change*. There was a photoshopped picture of Will in a dress and wearing girls' jewelry. It was a public page so anyone could see the picture and what was posted. I was positive Paul had created it.

I'd never seen kids gang up on anyone like this. All it took was one person, Paul, to get a group set up and then to rally everyone into thinking Will was a horrible person. I scrolled the list of people who had joined the group. Since our school is so small, we know everyone. There were so many names on the list that I didn't recognize that it was clear many of the people who had joined probably didn't even know Will personally but knew someone from our school who had invited them to join the group.

***Phil Larson: MOOOOOOOOOVING on up from animals to people Barnyard.***

***Sarah Wilks : Does it matter wht sex he is? Wuld n e one ever date him? LOL NO!!!***

***Steve Jones: ROFL Sarah right, well mybe a barn animal would date him MOOOOOOO***

***Amber Smith: OINK OINK***

***Stephanie Adkins: LOSER!!!!!!!***

***Chris Porter: STOP!! Will is a good guy. Knock it off!!!!!!! Jerks***

***Phil Larson: WA WA WA Chris, get a life. Chris when Will gets sex change, that mean u guys wuld b gay? LOL***

***Crystal Blanchet: ROFL Phil does it matter, gay r straight no one would ever date this loser, does he have any friends, beside chris but chris is loser too so WHO CARES***

***Phil Larson: Crystal, u r soooooo smart, I knw they can live in a barn 2gether LOL***

***Claire Newman: LOL who cares where he goes, JUST GO!!!!!!!!!!***

***Rebecca Kingman: LOZER!!!! I still cant believe that he put that pic up on his site of Olivia and him. There is NO WAY Olivia wuld ever hang with him. Only way is ybook club. Olivia being nice take pic for ybook and this lozer takes it way 2 far. WILL GO AWAY no one wnts u around here***

***Marc Brady: WOW becca tell us how u relly feel LOL group should be GO AWAY WILL***

***Rebecca Kingman: Marc u r wrong if Will went away we would miss him NOT!!!!!!!!!!!!!***

***Sarah Wilks: LOL becca u crack me up. Will see what u have done, no one likes u***

I scrolled down farther to see if the tone of the messages changed, but the more I read, the more my stomach twisted. I'd heard about groups like this before, but I'd never seen one. The fact that so many kids would participate shocked me. What was most disturbing was that there were very few comments from kids disagreeing with the activity and comments on the page. Even my friends were being mean.

I sat back in my chair, feeling helpless. I couldn't see any way that Will could fight back against these comments. Then I thought of the message I'd sent him telling him to take the picture down. Even the tone of *my* message felt harsh. Will thought I was really nice? I was sure he didn't feel the same after he read that.

For a moment I felt partly responsible for what was happening. I had taken that photo at the meeting. But Will shouldn't have used it on his *FriendPost* page, so he'd brought this on himself. I decided I shouldn't feel responsible for a choice he made.

My dad's voice broke my train of thought. "Liv, dinner's on."

"Be right there." I forced myself to leave the computer and go downstairs, but I had no appetite. I wondered where this would all end.

At school the next day the halls buzzed with talk about the group against Will. I was relieved to go into classes to focus on something else. After I left Latin I had to go upstairs for English. I walked through the back hall that passes a huge storage closet and two rooms that were currently empty. As soon as I walked through the door to the next hall I heard commotion and looked beyond the stairwell. A group of twenty or so kids were gathered in a circle staring into the middle. In between taunts and loud laughter from the kids on the outside, I heard Paul's voice.

"Come on, just a little red to accent your wardrobe, *Will*ma." He drew out the name then laughed.

I couldn't imagine who he was talking to. There wasn't a Wilma in our high school or anyone with that nickname, either. I hurried up the stairs and by the third step I was above the action. I gasped and goosebumps jumped out on my skin. Will was cowering in the middle of the circle and Paul had him in a headlock while he smeared lipstick on Will's mouth. A bright pink and silver boa was draped around Will's neck and his eye lids were caked with bright blue eye shadow.

After a moment Paul loosened his hold and pushed Will away. Will stumbled a few feet before he regained his footing. At the same time, the door from the hallway slammed open and Chris shoved his way through the tight group of kids.

Paul stalked toward Will and stabbed Will's chest with his finger. "Stay away from my girlfriend, Barnyard, or you'll get worse than this. Got it?"

Will didn't respond.

"Leave him alone!" Chris shouted as he worked to get to the front.

Ignoring Chris, Paul grabbed the front of Will's shirt and thrust his face close to Will's. "Do you understand?" he growled.

Will's eyes, wide with fear, stared up at Paul. He nodded. Then as if suddenly sensing I was there, Will looked up at me with an expression of disbelief.

Paul shoved Will one more time then turned to take his books from a friend standing nearby. "That profile picture better change tonight, Barnyard. And if you're lucky, I won't be seeing you around."

Like the Red Sea parting, the group of students opened up to allow Paul through. Then they headed in different directions, some going out the door and some passing me on the stairs.

I was frozen in place. Will stared up at me.

"Come on, man," Chris said as he removed the boa. "Let's go get this stuff off you."

Will shrugged him off but never stopped staring at me. "No, I'll go home and take care of it." Then he stepped closer to the stairs and grabbed the banister in front of me. "Do you still think I rock? Is that why you just watched what was happening?" There was an unusual sneer in his voice that made me uncomfortable.

My breath caught in my throat, but I finally found my voice. "There wasn't anything I could do."

"You could have stood up for me." The dull look in his eyes frightened me. He dropped his hands and took a step back. "I thought you were my friend."

I didn't know how to answer, so I just stared. An apology stuck to the tip of my tongue, but for some reason I couldn't say it. Will wiped his shirt sleeve across his mouth, streaking the cuff with bright red. The silence in the stairwell was punctuated by his heavy breathing. Finally, he gathered his books that had been strewn on the floor and headed for the door. When he reached for the handle, I came back down the three steps to be at his level. At the same time Chris crossed to the door, opened it and held it for Will.

"Take the profile picture down, Will," I pleaded. "That will solve the problem."

He nodded slightly. "Thanks for the advice, but I've decided how to solve my problem."

When he started through the door, I stopped him. "Will you be at the yearbook meeting in the morning?"

He hesitated then surprised me by smiling. "Oh, yeah, I'll be there."

The knot in my stomach loosened. "Cool. See you then."

He and Chris left, and the door squeaked closed behind them. I hoped he would take my advice and take the profile picture down so Paul would get off his back. I continued up the deserted stairs, knowing I'd be late to English. It didn't matter. I was never late, so I knew Mrs. Gardner would cut me some slack this time. I definitely needed to be distracted after this scene.

That night I checked Will's profile on *FriendPost*. For some reason I expected he'd leave the picture up just to show Paul. To my relief, it was gone. In its place was a picture of a gravestone. That seemed bizarre, but at least he'd listened to me. I posted a message for him that I'd see him at the meeting, but he never responded to it.

Next I checked the group page Paul had started against Will. Someone had posted pictures from this afternoon that they must have taken on a cell phone. They showed Will wearing the boa, lipstick and eye shadow. This had now gone too far, so I sent Paul a text message.

***Will took down the profile pic. Remove the group page. It's cruel.***

**Ur sticking up for him?**

***Back off. You made your point. Take down the group or there will never be a chance of getting back together.***

I had no intention of ever going out with Paul again, but at this point I just wanted him to take that down, so I let him believe what he would. He didn't respond.

Before I went to bed that night I checked and the group was still active with many more postings. Disgusted, I was ready to log off when my attention was caught by a post that sent a chill through me.

***Will Barnard: Thanks everyone. I know I rock***

I stared at his words; words that I had used during our yearbook meeting to compliment his photographs. When I'd given him the compliment I was sincere. Now, twice, he'd brought them up in the middle of horrible situations. My heart sank. Despite all of the vicious things people were saying about him, I hoped I could make him believe that I meant what I said.

The next morning I got up extra early. I wanted to be the first one to the yearbook meeting and hoped Will would be early, too. While tossing and turning during the night I decided I did owe him an apology, and it would be the first thing I'd say when I saw him.

Since my dad teaches 7th and 8th grade math at my school, he agreed to go in early so I'd have a ride. When we arrived the janitor's truck and a couple other teachers' cars were in the parking lot, but Mr. Rice's car wasn't. Will walked to school, so I looked down the street to see if he was coming. It was 6:45 a.m. and the janitor opened the building at 6:00 a.m. so it was possible Will had already arrived.

I walked in with Dad, but then he went off to the staff room and I went to my locker first to put away the books I wouldn't need for morning classes. On my way back down the hall I passed Mr. Locke, the guidance counselor.

"Good morning, Olivia," he greeted me. "You're here early."

"Yearbook meeting. We're on deadline, you know." I thought that sounded pretty cool.

"I saw Will Barnard come in earlier. Isn't he on the staff, too?"

My heart leaped. I'd get my chance to talk with him alone. "Yes, he's a photographer."

"Nice young man," Mr. Locke said. "Mr. Rice here yet?"

"No, but I can get things laid out so we're ready to go when everyone gets in."

He smiled. "True leader qualities."

"Thanks," I said. Mr. Locke was one of my favorite people at the school because he seemed to have a compliment for everyone.

He went on his way and I continued to the technology wing. One of the first things Mr. Roy, the janitor, does when he arrives at school is to open all classroom doors, so I knew I'd be able to get in. The hall lights were on timed sensors and came on as I entered each section, which told me no one had come down the hall in at least the last fifteen minutes. I wondered where Will was. A chill shot up my spine. I'd never been the first one in this part of the building in the morning and the quiet was eerie. I increased my pace so I could get to Mr. Rice's room and distract myself with work.

I hurried into the room and flipped the light switch. The fluorescent lights flickered then filled the room with brightness. A strange noise came from the computer area around the corner. I stopped and stood like a statue, listening for the sound.

"Will?" There was no answer. "Will, are you in here?"

When I heard what seemed like a muffled gagging sound, I darted out of the room and stood outside the door. My heart pounded and fear crawled along my skin.

I stared into the room for a minute wondering if my mind was playing tricks on me because I was jumpy about being here alone. From the hall I couldn't hear any sounds, but I wasn't particularly anxious to go back in, either. Before I could convince myself to enter again, I recognized Mr. Rice's whistling down at the other end of the hall. I took a deep breath and let it out slowly, hoping he wouldn't see that I was shaking. I felt like such a baby for being scared.

He came around the corner and stopped whistling when he saw me.

"Hi, Mr. Rice." My voice echoed down the big, empty hall.

"You're quite the early bird," he said. There was a lightness in his step that made it seem like he was always happy.

"Tweet! Tweet!" I joked, finally feeling more relaxed.

"Go ahead and go in," he said as he got closer.

I considered telling him I thought I'd heard a noise from the computer area then decided not to. I felt safer with him here, and it was probably just my imagination anyway. I went into the room and set my books on a desk. Then I went to the back of the room and pulled out the boxes of yearbook materials we needed to work on this morning.

Mr. Locke started whistling a tune I didn't recognize and moved about the room laying out materials we'd need. He was back by his desk when I heard the same sound I'd heard when I was in the room alone, only it was a little louder this time. He snapped his head up and looked at me.

"Did you say something?" he asked.

"No, but when I arrived I heard that same sound and thought it was my imagination. That's why I waited for you in the hall." I pointed toward the computer section of the room that was behind a half wall. "I think it came from over there."

He furrowed his brows then crossed the room quickly. I took a few steps in that direction but decided to wait for him to tell me there was nothing there. He stepped to the other side of the half wall and I heard him gasp.

"Will. Will?" His voice rose and I could hear fear in it. "Will!"

My heart battered my rib cage, and I raced toward him. "Mr. Rice, what's wrong?"

"Stop, Olivia! Don't come over here," he ordered.

I lurched to a stop, every inch of me shaking. "What's wrong?"

"Run to the office and get Mrs. Gardner if she's here. If she's not, I know I saw Mr. Locke. Tell him to call 9-1-1. We have a student down."

"Will?" I cried.

"Go, Olivia. Now."

I raced down the hall, my legs feeling more like rubber the faster I ran. When I got to the office, Mrs. Gardner, the principal, and Mr. Locke were standing at the counter looking at some papers. Their heads snapped up when I ran into the office.

"Olivia? What's wrong?" Mrs. Gardner asked.

"Mr. Rice said to call 9-1-1. It's Will Barnard. Something's wrong."

Mrs. Gardner went to the secretary's desk and picked up the phone while Mr. Locke hurried around the counter. "Where is he?"

I gasped, trying to catch my breath. "In Mr. Rice's room. Please tell them to hurry. I think it must be bad."

Mr. Locke raced from the office. His feet pounded down the hallway.

Mrs. Gardner held the phone to her ear. "What's wrong with him?"

"I don't know. I didn't see him. He was on the side by the computers." My voice shook, but I couldn't help it.

The 9-1-1 dispatcher must have answered because Mrs. Gardner very calmly relayed the information to them. When she hung up the phone, she used the walkie-talkie to call Mr. Roy to the office.

"Is your dad here, Olivia?"

"Yes."

She picked up the phone again and punched in some numbers. The intercom beeped, indicating an all call through the building. "Rick Dawson please come to the office immediately."

She hung up the phone then came and led me to one of the chairs by the big window that overlooks the halls.

"Olivia, sit right here and wait for your dad." I sat, relieved to be off my wobbly legs. "The two of you should wait for me to come back. I'm going to have Mr. Roy wait by the front doors to show the ambulance crew to Mr. Rice's room. It's important that you stay here. Do you understand me?"

Mrs. Gardner's calmness helped calm me, but still my muscles quivered like I was freezing. "Y-y-yes," I stammered. I couldn't believe this was happening. I wanted to know what was wrong with Will.

She met Mr. Roy in the hall, spoke with him for a moment, then she hurried toward the technology wing, as well. I stared out the window, silently pleading for the ambulance crew to hurry.

My dad came down the hall next, and I saw Mr. Roy speak to him and point toward me. Dad hurried into the office, and I jumped up to hug him.

"Olivia, what happened?" he asked.

I explained what little I knew as he took me back to the chairs. He put his arm around me and we waited. Finally there were sirens in the distance, and a few minutes later a woman carrying a big bag and a man and another woman pushing a stretcher barged through the door and followed Mr. Roy down the hall. The teachers who were in the building came into the office. First they looked for Mrs. Gardner and Mr. Locke, and when they couldn't find them, they came back out and questioned Dad about what was happening.

Miss Courchesne, a teacher I'd had for history classes in middle school, sat down next to me.

"You okay?"

I shrugged, not trusting my voice.

No one wanted to leave the office while we waited to hear about Will. The teachers talked in low voices about yesterday's big tennis match between our team and Wentworth Central and the mid-term exam schedule and who was going away over April vacation. I knew they were trying to make me feel better by acting calm and natural, but my knees ached from the tension. I counted the floor tiles between the door and the counter. Anything was better than wondering why it was taking so long for them to come out with Will.

All at once their voices hushed, and I glanced up to see them staring out the big office window. I swiveled in my chair in time to see the ambulance crew and Mr. Rice rushing up the hall with Will on the stretcher, his arms strapped to his sides. A silver and green oxygen tank laid between his feet. Clear tubing ran from the tank to an inflatable bag that a technician squeezed every few seconds to force air into the big mask that covered Will's face. Medical wires stretched from him to the monitor they'd balanced on his legs.

Mrs. Gardner and Mr. Locke followed just behind them, deep in conversation and with grim expressions that scared me. We kids always joked that we wouldn't want to play poker with Mrs. Gardner because her expressions never change. I was seeing a very different principal now.

"Hold up, Sally," the male technician ordered the woman pulling the stretcher. "Give me a second to check the vitals."

They stopped just inside the school entrance. The technician hunched over Will's body to study the machines. "Come on, kid, you're not doing this to us," he growled. He moved a couple of wires then demanded, "Let's go," and they picked up their pace toward the front door.

I couldn't tell if Will's eyes were open or closed, but what skin I could see on his face was chalk white. The ambulance technicians kept talking to him, but he didn't respond. I'd only seen scenes like this on television and in movies. A huge lump filled my throat so it was hard take a full breath. Mr. Roy ran ahead of them and blocked open the front doors so they could roll the stretcher through.

"Olivia, don't watch." Dad grabbed my shoulders and tried to force me to turn around, but it was too late. I'd already done my own assessment of the situation.

My palms hurt. When I unclenched my fists I saw my fingernails had left deep indentations in my skin.

"He looks dead!" I cried.

Now Dad forced me to turn so I couldn't see out the window. "I'm sure he's not. All of those machines often make things look worse than they are."

"I hate this!" Even though saying that made me sound like a little kid, it was how I felt.

"He's being taken care of, Liv. We'll hope for the best for him."

The commotion in the hall faded as they hurried through the front door. A minute later the sirens wailed as the ambulance raced toward the hospital.

Mrs. Gardner came into the office and glanced at the teachers and me before she entered her office, closing the door behind her. Mr. Locke and Mr. Rice came in shortly after, looking worried. I wanted to stand up but my legs felt too weak to hold me.

"Is Will okay?" I asked.

Mr. Rice kneeled in front of me and Mr. Locke sat in the chair on the other side. "He's still alive," Mr. Rice said.

My heart jumped. "Still *alive*? Why wouldn't he be?" I looked back and forth between them to try and read their expressions.

"Mrs. Gardner is calling Will's parents right now so they can meet him at the hospital. Then she'll talk with you and your dad before you go home."

"Why do I have to go home? What's going on?"

"Let's wait for Mrs. Gardner to get off the phone then we'll go into her office to talk, okay?"

I slumped back in the chair knowing I wouldn't like what they were going to tell me. Why else would I have to go home?

Other teachers and office workers filed in and Mr. Locke ushered them to the hallway, closing the office door so I couldn't hear as he spoke with them. It seemed like forever before Mrs. Gardner came out.

She looked at Mr. Rice and Mr. Locke. "Jim. Dan. I'd like you to come in with Olivia, too," she said. Dad took hold of my arm and we entered. Mrs. Gardner and Dad sat in chairs next to me and Mr. Locke and Mr. Rice stood in front of her desk.

Mrs. Gardner put her hand on my arm. "Olivia, I'm sure you're very worried about what's going on."

"I just want to know if Will is going to be all right."

Mrs. Gardner glanced away for a moment. When she looked back at me I could see pain in her eyes. "We honestly don't know."

I pressed my fingers to my lips, hoping I wouldn't cry. "What happened? Is he sick?"

She shook her head. "No, not really." She seemed to think carefully before speaking. "Olivia, Will typed a message on the computer." She paused. "To you."

"To me? What did it say?"

"It said, 'Olivia, do I still rock?' Do you know why he wrote that?"

Tears welled in my eyes then before I could stop them, I was choking back sobs. I clutched my stomach, willing the sudden pain to go away.

"Are–are you t-telling me that W-Will tried to k-kill himself?" I cried, hating the question.

Mrs. Gardner took my hands and held them. "That's how it appears." She gave me a minute for that to soak in then added, "Now we need to know why."

# ISSC
# Internet Safety Savviness Challenge

See how many of these questions you can answer then check your answers with the answer key at the end.

1. Define cyber bullying.

2. If you know a friend is being cyber bullied, is there anything you can do to help. If yes, what?

3. If there is an online "group" or "page" making fun of one of your friends, should you go onto that group and post comments? Explain.

4. You talk to some of your friends about a cyber bullying situation and you feel that the situation is serious. Who should you talk to next and why?

5. What are some warning signs that you might notice if a friend is in danger or not feeling good about him/herself?

6. What are steps that you can take to prevent yourself from becoming a victim of cyber bullying?

# What Can We Learn?

The events that unfolded in this chapter, although fictional, are occurring EVERY day in communities all across this country. We can't think of bullying as a "rite of passage" for kids or assume that insults and sarcasm are the result of kids being kids. Today we have to look into things a little deeper and with an open mind that it might not be just picking on a kid. These actions may be destroying a child's self-esteem and causing major, life-long problems.

Technology has taken bullying to a new level and has made it a bigger concern than it already was. Bullying is not going to go away. It has been around forever and will remain part of our lives. However we deal with it, it must change.

Let's review and see what went wrong in our fictional scenario and what action could have been taken to help prevent this situation. Olivia realized that Will was being bullied and harassed by Paul and other students when they were tripping him, calling him names and making animal sounds at him. All of this behavior can take a toll on a person. Victims of bullying can suffer from emotional issues that are very difficult to deal with such as lower self-esteem and self-worth, isolation, lack of confidence, lack of trust in others and a feeling of hopelessness.

Olivia could have approached a teacher or counselor and explained the situation. There is a major difference between being a tattletale and being a mature person doing the right thing. Telling a teacher or adult at the school that a student is being bullied is one of the most mature decisions a student can make.

In the following exchange with a friend, Olivia did something that took courage.

***Olivia Dawson: everybody's getting worked up over a profile picture and hassling me***

***Shawna Richards: tell him to take it down***

***Olivia Dawson: why? He likes it.***

***Madonna McCartney: whatevah***

***Olivia Dawson: yeah whatever lol***

When confronted with this type of pressure from friends, Olivia could have become very upset and started yelling or making up excuses or lies as a way to explain why Will would put this picture up. Instead, she chose to not react, which was a proper response. Unfortunately, Olivia missed out on an opportunity to make one more appropriate decision; she could have reached out to Will and discussed this matter directly with him. If she had done this, Will could have learned what people were saying, perhaps better understood the impact on someone else and then maybe reconsidered using the picture. Chris, the one true friend Will had throughout this ordeal, tried to help but unfortunately he decided on the typical route to resolve the issue. Instead of approaching an adult for help, Chris went to another student, Olivia.

Teens want to help each other and not involve adults in every aspect of their lives, but there are times when adults NEED to be involved. This was one of those times. The wall postings, Will skipping classes, not eating in the café and missing yearbook meetings were all signs that Chris picked up on, and he was correct to follow his gut instincts to tell someone. Telling Olivia about Will's problems was a brave and mature move, but if Chris had taken it one step further and talked to his parents, Will's parents, or another trusted adult, some intervention may have taken place to help Will.

Once Olivia saw the postings that had concerned Chris, she should have done the same thing. Kids are not going to, nor are they expected to, tell adults every time a kid is having a bad day or some kids make a comment about being angry or upset. However, as in this situation, if a child believes there is a serious problem, it should be looked into and dealt with. If the situation turns out to be nothing, consider it a learning experience, but if it *is* something, seeking help for the friend as early as possible can prevent some very tragic incidents from occurring.

***"I'm getting hassled about your profile picture. Change it."***

Is this how a friend would deal with this situation, especially if he/she knows the other person is dealing with some emotional issues? Olivia could have handled this so much better. For example, rather than just saying that she was "getting hassled", she should have communicated better and explained how the posted picture was creating a negative situation for both of them. By being specific, perhaps Will would have understood that removing the picture could improve the situation for both of them.

Almost every social networking site has ways to report inappropriate behaviors and can usually be found on the site's home page. No one reported the site about Will; instead they contributed to the bullying. Chris tried to stop the mean comments but look what happened; people started insulting him. This behavior plays right into what the bully wants: more attention and more people involved. Chris could have reported the site as harassment and hoped that the hosting social networking site would take the page down.

When we are bullied at school it stinks. If we are being picked on we make it through the day, as miserable as it may be, and we go home. Our homes should be the safest place on Earth for us, and we should never feel uncomfortable there. Since the introduction of the Internet and the cyber world, our homes are no longer the isolated places they used to be. Because of the Internet, when people are bullied, like Will, home does not feel safe.

Put yourself in Will's shoes. The bullying that he faced was occurring everywhere he went, and in his mind there did not seem to be an end in sight or a safe place to get away from it. His stress never stopped. The Internet adds this new opportunity for bullying: it can be never-ending and constant. Unfortunately, Will reached his limit. No one should have to live through this.

There are great features to today's social networking and not everything that happens on these sites is bad. When used properly, they can be a good way to communicate and stay connected with people; however, when these networking sites are abused or used in a way they are not meant to be used, situations like Will's can, and do, occur.

Getting the right people involved as early as possible may be one of the best approaches to dealing with these situations. Hoping the bullies or the situation will go away, looking the other way, laughing and supporting the bully, or just doing nothing can lead to a very tragic and sad situation.

Get involved and be strong if you think someone is in this type of danger. It is not only the right thing to do, you may save a life.

# ISSC Answer Key

1. Cyber bullying is when a person is tormented, threatened, harassed, humiliated, embarrassed or targeted by another using the Internet, mobile phone or another type of digital technology.

2. Yes, there are ways you can help if you know a friend is being cyber bullied. First talk to your friends and let them know they are not alone and people care about them. Tell a trusted adult about the situation (parents, teachers, counselors, clergy and law enforcement). You can also "report" the situation to the web site who is hosting the group and advise them of the situation. Being a good friend is crucial to getting help.

3. No. If there is an online "group" or "page" making fun of one of your friends, you should not go onto that group and post comments. Even if you go into a group or page and defend your friend, you are doing exactly what the bully wants, giving the situation attention. Ignore the temptation. Instead, follow the steps we have outlined in the above answer.

4. If you feel a bullying situation is getting serious, you should talk to a trusted adult. It is important to have trusted adults in your life for many reasons. One very important one is you can turn to them when you do not have an answer on how to help yourself or a friend. Their life experiences help in the decision making. Adults can help and need to be involved.

5. There are many warning signs that someone may be depressed or in danger of hurting himself. Isolation, anger, overly emotional or upset over little things, paranoid, a change in eating patterns, lack of interest in activities that normally are enjoyed or a person who is just not "themselves" are a few. Any of these signs or others should not be ignored. If you do not know how to help, talk to someone who can.

6. There are many steps you can take to avoid becoming a victim of cyber bullying. First, limit your exposure to social networking. Cyber

bullying is made easier when the bully can obtain your information and pictures. Next, avoid sending pictures and information over mobile devices. People can easily "forward" information and that can end up in a bully's hands. Talk to your friends and inform them not to share your picture or information. Finally, tell your friends you do not want your picture on their social networking site, don't allow people to "tag" or identify you in a picture and emphasize that no personal information about you should be shared on a friend's site.

# CONCLUSION

When we decided to write this book our intention was not to scare kids into thinking they're in danger every time they sign on-line. Our intention was not to have every parent frightened to the point they wanted to smash their computers and never allow their children to use technology again, even if they did think it a few times while reading. Our goal was to heighten everyone's awareness about safety.

Fear should not govern Internet use; education should. Technology is a great tool that is ever-changing and will continue to evolve with new ways to connect with the rest of the world. If we do not allow our children opportunities to learn or use any of these great resources, it could put them behind both educationally and socially. But, with that being said, if we don't learn how to use technology properly and safely, the results could potentially be tragic.

If this book makes even one child reconsider about posting a picture, makes a parent look into their children's on-line activities, or encourages a child to say no when a stranger talking to them on-line wants to meet, then the work we put into this project was well worth it.

New safety issues come to light constantly. If you want to remain current regarding the latest safety concerns with Internet and technology use, you can visit either:

www.InternetSafetyConcepts.com
or
www.R U in Danger.net

# Appendix A

# Searching a Computer For Web Sites Visited

As a parent it is very important to not only set up rules in regard to Internet and technology use but also to have the knowledge and tools to "occasionally" check to see what your children are doing. One way to keep tabs on your child's Internet activities is by checking the web browser "history". The web browser is the program that is used to surf the web. The most common browsers would be "Safari" for MAC users and "Internet Explorer" for the PC, but it is not limited to these two.

Each browser may be a little different; some may save sites for longer and some may even save search terms that were used in search engines. No matter what they save, every bit of information can be helpful. Below are some of the popular browsers that are being used. Listed with each browser are different ways that a parent can check the history.

***Safari*** In the web browser click "ctrl" and "h"

OR

Upper left hand corner of web browser is the icon for "show all bookmarks" (looks like an open book). When you open this it will show you "collections" and "bookmarks". Under "collections" is history.

OR

Upper right hand corner is an icon for "settings" (looks like a small gear)

In this drop down menu look for "history"

***Opera*** In the web browser click "ctrl" and "h"

OR

On the left side of the web browser screen locate the icon that looks like a clock. You click on this icon the history pane will open.

***Google Chrome***

In the web browser click "ctrl" and "h"

OR

Upper right hand of the web browser is an icon that looks like a wrench. This icon has a drop down menu for "customizing and control". In this drop down menu you will locate "history".

***Flock*** In the web browser click "ctrl" and "h"

OR

In the tool bar at the top of the web browser there is a tab called "History". When you click on this tab several options on what you can see for history appear.

***Mozilla Firefox*** In the web browser click "ctrl" and "h"

OR

In the tool bar at the top of the web browser there is a tab called "History". When you click on this tab several options on what you can see for history appear.

***Internet Explorer*** In the web browser click "ctrl" and "h"

**OR**

In the upper left hand area of the web browser there is a "star" icon. If you click on this star a side pane will open and show you the browsing history.

***Maxthon*** In the web browser click "ctrl" and "h"

**OR**

In the top of the web browser is a "click" icon. Clicking on this icon will open up the history tab

***Avant*** In the top of the browser there is a "clock" icon with a green arrow. Clicking on this icon

***Browser*** opens up the history pane.

OR

Upper right of the web browser has tabs. One of them is "view". Under "view" is a tab for history.

***Deepnet Explorer*** In the web browser click "ctrl" and "h"

OR

On the top task bar you "right click" your mouse. Choose "customize". Under "customize" find "history" icon (looks like clock with green arrow). Highlight this icon and then add to "current". The icon will then appear on your top task bar in the web browser for easy access.

# Appendix B

# How to Set Search Engines to a "Safer" Mode

A search engine is a web-based program that searches and retrieves information. Search engines are a huge resource offering information that can help with school work, finding activities for families to do, planning vacations, researching items before you purchase or just finding information to end an argument. Search engines are one of the great benefits of the Internet.

Almost every search engine on the web has a method for parents to set it up so when a child initiates a search they are not exposed to adult content. As most of us may have experienced, we go to a search engine and type in a search topic, not even thinking it could be a reference to something perverted or demented. But what do we see? Images and sites that we had no intention of viewing. Setting search engines to a safer mode will prevent most of these searches from taking the user to an inappropriate site.

Each search engine has a unique approach for changing the user settings and preferences. Listed below are step-by-step directions for you to follow to implement these settings for some of the most used search engine sites. (Web sites change from time to time. The guidelines below are current at the time this section is being written, April 2010. For more current information, visit my website: www.InternetSafetyConcepts.com. The latest updates will be listed there.)

1. **Google.com** - From the main screen, find "search settings" (usually in the upper right hand corner). When you click on "search settings" it will take you to the "Preferences" page. On this page you may choose the language you want to use but toward the bottom of

the page you will find “Safe Search Filtering”. Select “Use strict filtering (Filter both explicit text and explicit images)”. Save your preferences.

Google also offers “Locking Safe Search”. When you “Lock” the safe search you are making it more difficult for anyone to change your settings. To use this feature, go to the Google home page, click on “Settings” and find “Lock Safe Search” (You will need a Google account to use this feature). If you have separate user accounts for different family members, each account will have to be separately set up and “locked”.

2. **Yahoo.com** - On the Yahoo home page you will see the box that allows you to type in what you are searching for. Next to this box is a box called “Web Search”. Do not type anything in the search box but do click on the “Web Search” box. This brings you to a search feature in Yahoo. Over the search box is a button called “Menu”. When you click on “Menu” a drop down list appears. Click on “Preferences”. When the “preferences” window opens look for “SafeSearch”. Click edit which will allow you access to change the settings. Check “SafeSearch lock”. In using this lock, this will filter out adult web search results including video and image searches. This will apply to any user on that computer who uses Yahoo.

3. **Ask.com** - From the main screen find “Settings”. Click “Content Filtering”. When this screen opens up select “Filter my search results”. This will limit adult orientated content.

4. **Bing.com** - From the main screen, “Preferences”. Once in “preferences” locate “SafeSearch”. Here is where you can select the level of restriction you would like (Strict, Moderate or Off). Be sure to save preferences prior to exiting.

5. **Dogpile.com** - From the main screen select "Preferences". In "Preferences" under "Search Filter" you can choose your restriction level (None, Moderate or Heavy)

6. **Altavista.com** - From the main page select "Settings". On the "settings" page locate "Family Filter". On this page you can "choose your Family Filter preference". (All, multimedia only, or none)

   Obviously there are many more search engines that are out on the Internet, but most of them will have features that parents can use to help prevent children from being exposed to the wrong information. Almost every search engine also states that, although their filters are good and they are making an effort to keep children safe, no filter is 100% guaranteed. This is why parents should have a lot of tools in their "parent toolbox".

Setting the search engine to a safer mode is a great resource but should not be the only one. With these settings, the use of parental controls and openly talking with our children about the dangers that could occur if we are not careful, it is a good foundation for keeping all of us from being exposed to content that we never should see on the Internet.

# GLOSSARY

## COMMON TERMS ASSOCIATED WITH INTERNET AND COMPUTER USE

Anti-Virus - A software program designed to identify and protect computers from known viruses.

Attachment - A file that is sent with an e-mail or message. Almost any type of file may be an attachment, songs, data, spreadsheets and photos. Most Internet Service Providers will have limits on the size of attachments that can be sent

Bookmark – Used with an Internet browser, a bookmark saves the link to a website. When you add a bookmark to your list it allows easy access to go to that site again in the future. Some browsers will refer to bookmarks as "favorites"

Browser – Software program that allows a user to navigate the World Wide Web

Buddy List - A contact list of screen names. Buddy lists are most commonly found in instant messaging programs but can also be found in cell phone programs, game rooms and some social networking sites.

Chat – Form of on-line communication. Chat can occur in Chat Rooms, Game Rooms and other media. Chat is usually done in "real time" (as soon as you type something and send it, it can be read immediately by anyone using the same program)

Cookies – Piece of text stored on a user's computer. Cookies are saved when a user visits a web site using a browser. Cookies help speed up opening websites and can assist parents in finding out what sites their children are visiting on the Internet.

Cyber Sex - Virtual sex encounter. While using the Internet or mobile technology, two or more people communicate with sexually explicit comments. Cyber sex can occur in instant messaging, one on one, or in chat or game rooms involving numerous people at the same time.

Download – Copying a file from the Internet to your own computer or personal device. Files that can be downloaded can range from data files, movies, software programs and music.

E-Mail (Electronic Mail) - Messages that are sent from a user's computer or mobile technology to one or multiple recipients using the Internet as a delivery device.

Emoticon – Text or symbols. Often used in instant messaging, chat or e-mails to help show emotions. For example :-) will represent a smiley face.

File – Computer Data that can come in a wide variety (Documents, programs, applications and other computer data)

Firewall – Security that limits the exposure of a computer to a network or the Internet. Uses both hardware and software programs. Firewalls assist in preventing people from gaining access into other people's computer system.

Hacker – Someone who "breaks" into another person's computer. Usually hackers will use software programs to gain access to another's computer for the purpose of harassment, theft or stealing one's identity

Hard Drive – Device used by computers to store data. Hard drives can be found inside a computer or can be "external" (connected to computer via USB or other cables)

Hardware – Electrical and physical components of a computer system. Hard drives, CD readers/writers and processors are all examples of computer hardware.

History – Found in the Internet browser, the history folder stores information on web sites that were visited by users. Most browser histories will show date and time a site was visited.

Instant Messaging (IM) – Communication between two people in a private one to one conversation that occurs over the Internet or mobile technology. Most IM programs will run in "real time", in other words, they're taking place at that moment.

Internet – A global computer network system.

Internet service provider (ISP) A company that provides Internet access

Link – Usually found on web sites or included in messages, a link is a command that directs to another location. If you are on a web site and a link states "to go to InternetSafetyConcepts.com web site click here", when you click on it you will be brought to that web site.

Maximize – When a program is "maximized" it takes up the entire computer screen. In most programs there is a button in the upper right hand corner that allows a user to "maximize" the screen.

Minimize – When a program is minimized it disappears from the screen. Most programs that are "minimized" will still appear active in the task bar portion of the screen. To bring the programs back to the screen, user simply has to click on the program in the task bar. It is often shown as a box partially within a box.

Operating System (OS) - OS is an interface between hardware, software and the user of the machine. OS allows programs to run.

Password – Secret word, phrase, numbers, symbols or letters that allow a user access to protected systems or programs.

Program - Instructions that a computer interprets so it knows what to do, "run" or execute. Programs contain codes that are read and executed by the operating system.

Real Time - When it comes to chatting or instant messaging, real time means as soon as you hit "enter" or "send" that information is immediately viewable to the person/persons you sent it to. Once a message is sent "real time" you can not cancel or take it back.

Screen Name - Name used when using on-line programs, games, instant messaging, chat and others.

Search Engine - a web-based program that searches and retrieves information from the Internet. (See Appendix for a sample of popular search engines.)

Spyware – Virus or "malware" that is installed and collects information. Viruses and malware can be put on to a computer by several different means. One is by downloading an infected file, another is through an attachment in an e-mail purposely installed by a hacker or stalker and yet another is from transferring files from one system to another. The user whose computer is infected by spyware may not even know the computer is infected. Spyware can run without a person's knowledge

Taskbar - A bar, usually at the bottom of the screen of a Windows program, that shows what programs are currently running. Other things found in the taskbar could be the "start" button, clock and calendar.

Upload – Act of transferring a file from your computer to another computer or device. Files can be uploaded to either a computer system or Internet-based sites/programs.

URL – "Uniform Resource Locator". Words, letters and numbers that identify a web site's address.

## COMMONLY USED ACRONYMS IN CHAT, E-MAILS and INSTANT MESSAGING

In the world of texting, e-mails, chatting and instant messaging, a lot of acronyms are used on a regular basis. Some are common, but some may be hard to decipher or have multiple meanings. Below is a list of some about which families should be aware. New acronyms are always surfacing and a good way to keep up is to occasionally search the web for new IM/Text acronyms.

| | |
|---|---|
| 143 | I Love You |
| 2moro | Tomorrow |
| 2nite | Tonight |
| 4eva | Forever |
| 4ever | Forever |
| 4Q | F*** You |
| 831 | I Love You |
| ?^ | What Up? or Hook Up? |
| A/S/L | Age/Sex/Location |
| Addy | Address or College slang for ADHD medications |
| AITR | Adult in the Room |
| AML | All My Love |
| B4 | Before |
| B4N | Bye For Now |
| BAK | Back At Keyboard or Bad Ass Kid or Back |
| BB4N | Bye Bye For Now |

| | |
|---|---|
| BF | Best Friend or Boy Friend |
| BRB | Be Right Back |
| BRT | Be Right There |
| BTDT | Been There Done That |
| BTW | By The Way |
| CICYHW | Can I Copy Your Homework |
| CUL8R | See You Later |
| CUZ | Because |
| CYT | Cute Young Thing or Check You Tags or See You Tomorrow |
| DL | Down Low – Download |
| DOC | Drug Of Choice or Doctor or Department of Corrections |
| DURS | Damn You Are Sexy |
| DULM | Do You Love Me |
| EFFIN | F***ing |
| F2F | Face To Face or In Person |
| FB | Facebook or F***buddy |
| FOAF | Friend Of A Friend |
| FU | F*** You |
| FU2 | F*** You Too |
| FYEO | For Your Eyes Only |
| G2G | Got To Go |

GAL Girl or Get A Life

GMAB Give Me A Break

GR8 Great

GTFO Get The F*** Out

GTG Got To Go

GTH Go To Hell

HU Hook Up or Huh (Confusing)

I <3 U I Love You

IDK I Don't Know

ILU I Love You

ILY I Love You

IM Instant Message

IMRU I Am Are You

IMS I Am Sorry or Irritable Male Syndrome

IPN I'm Posting Naked

ISO In Search Of

ITA I Totally Agree

IWSN I Want Sex Now

IYQ I Like You

JDI Just Do It

JK Just Kidding

JO Jerking Off

| | |
|---|---|
| KFY | Kisses For You |
| KITTY | Code Word For Vagina |
| L8R | Later |
| LH6 | Let's Have Sex |
| LMAO | Laughing My Ass Off |
| LMIRL | Let's Meet In Real Life |
| LOL | Laughing Out Loud |
| LUSM | Love You So Much |
| LY | Love You |
| LYB | Love Your Body or Love You Babe or Love You Bye |
| LYL | Love You Lots or Love You Later |
| M4W | Men For Women |
| MSG | Message |
| N1 | Nice One |
| NBD | No Big Deal |
| NE1 | Anyone |
| NFS | Need For Sex or Need For Speed or No F***ing SH*T |
| NIFOC | Naked In Front Of Computer |
| NSA | No Strings Attached |
| NW | No Worries or No Way |
| OIC | Oh I See |

| | |
|---|---|
| OMG | Oh My God or Oh My Gosh |
| OTC | Over The Counter of Off The Chain |
| P911 | Phrase used to advise a parent is in room or watching |
| PCM | Please Call Me |
| PHAT | Pretty Hot And Tempting |
| PITA | Pain In The Ass |
| PIX | Picture |
| PLS | Please |
| PLZ | Please |
| PM | Private Message |
| POS | Parent Over Shoulder or Piece Of SH*T |
| ROTFL | Rolling On The Floor Laughing |
| ROTFLMAO | Rolling On The Floof Laughing My Ass Off |
| RU | Are You |
| S2R | Send To Receive |
| SNAFU | Situation Normal All F***ed Up |
| TDTM | Talk Dirty To Me |
| THX | Thanks |
| TLK2UL8R | Talk To You Later |
| TM | Trust Me or Text Me or Text Message or Trouble Maker |
| TOY | Thinking Of You |

| | |
|---|---|
| TTFN | Ta Ta For Now |
| UOK | You Ok |
| W8 | Wait |
| WTG | Way To Go |
| WYWH | Wish You Were Here |
| XTC | Ecstasy |
| Y | Yes or Why |
| YGM | You Go Me or You Got Mail |
| YW | You're Welcome or Young Woman or Yeah Whatever |